Writing User Documentation
(Second Edition)

Hints For Document Writers:
Types Of User Documents
Writing Style Suggestions
Organizing User Manuals
PDF, eBook, HTML Help
Procedure Formatting
Graphics, E-Learning

Clues, Ideas, And Possibilities For
A Beginning Technical Writer
To Improve User Documents

Don G. Miller

Copyright © 2009 Don G. Miller

EAN-13: 9781448632824
ISBN-10: 144863282X

Subject Headings:
User Documentation
User Manuals
Technical Writing
E-Learning

Trademarks

About The Author

Don Miller is a freelance technical writer with over 25 years of writing experience in electronics and electromechanical equipment. He has written, edited, organized, and published user documentation for Audiometers, Vision Testers, Semiconductor Test Equipment, Disk Test Equipment, Probe Card Analyzers, and industrial software applications and interfaces.

He has also written and self-published two full-color books of Arizona images, using Photoshop to enhance the images. Additionally, he has published a book on *Self-Publishing for the Thrifty*.

The author may be contacted by the following methods:
Don Miller
PO Box 1403
Payson, AZ 85547

Or

E-mail = Don@RimThoughts.com

The author is always looking for suggestions that might improve this book, so please feel free to contact him with your suggestions.

If using e-mail, please include "Basics of User Documentation" in the subject line to reduce his confusion.

DISCLAIMER

There are products mentioned in this book the author has used. There is no intent to defame or endorse any product. The information in this book is not guaranteed to be completely accurate and is provided "as is". The author accepts no liability or responsibility as to the use or misuse of this information.

Some Conventions

NOTE: Provides additional information about the topic.

TIP: Indicates a suggestion or tip that may be helpful.

Author's Aside
Asides are the author's personal observations, which may or may not be of any real value.

Contents

6   Contents

Contents 7

Appendixes

Preface

This book is the second edition of "Writing User Documentation". I received a few complaints about the first edition and after thinking about those complaints, I decided to expand and reorganize the book. I considered using a new title but decided that was not the best solution.

This updated edition includes screen recorder ideas, how and why we learn, how to work with graphics to achieve the best results, and how to be sure the PDF document you publish is what an end user sees.

This book's primary focus is on semiconductor testers that use proprietary user interfaces. It is written to show a writer how to organize a manual based on the tester's main menu structure. This book also describes how to write procedures so a person who has no real experience can safely perform those procedures.

It should be noted that each type of user documentation requires a different viewpoint as well as a slightly different set of skills and abilities. Even though this book's focus is on User Manuals and application HTML help, most of the ideas and suggestions can be applied to e-learning, eBooks, and other types of documentation.

Author's Rant

When I started writing technical manuals and user documentation, way back in the dark ages, technical writing was more about educating the user than about entertaining the user. Today, most manuals appear to look more like comic books than textbooks.

This means that user documentation, and especially manuals, are written to entertain rather than to inform. Most manuals appear to be written for the inattentive reader or a reader that has no real desire to learn. Today, manuals have just enough information to allow the user to operate the product, but most manuals no longer answer the "Why?" question.

Is this because no one expects a mere user to understand how the product operates or why it operates or what it really does? The user is just supposed to press this button, wait for something to happen but never wonder why or even how that something was done.

Manuals are now being written as if the user has no interest or curiosity and should just be an automaton so when this light goes on, press this button, when the buzzer sounds, place something here or there or somewhere. Why are we dumbing down the user? Should we not expect more from them?

Manuals should invite the user to learn more. Manuals should encourage the user to ask questions, to look for why things work, and to wonder how things can be made to work better or smoother or faster.

While including videos and expecting the user to watch is useful, not engaging the user and requiring some action on that user's part is as oversight. And not including more information about what the product does, how it does it, and why it does it is negligent, and in my opinion, a result of writer's laziness or lack of product knowledge. Too much user documentation is directed toward telling without engaging the user.

Some of the most useful feedback comes from users that wonder why the product works this way instead of that, why not add this feature, or why not do things this way instead? But these musings are not encouraged if the manual fails to go beyond the simple description of how to operate the product.

Users learn through three methods: listening, watching, and doing. Watching a video without requiring the user to participate will help a few users, hearing the information will help a few more users, but making the user perform the required steps, even in simulation, promotes a better learning experience for the majority of users.

I have ranted enough, but you get the idea. All I am really suggesting is try to educate, engage, and inform your users. Try to stimulate their curiosity, answer a few unnecessary questions, poke at their wonder, and avoid writing as if the user is too ignorant to understand the product. This type writing only encourages users to ask more dumb questions instead of trying to reason out the answer for themselves or even take the time to look for that answer.

Don

Chapter 1

Technical Writing Overview

Early History

I have no idea when the first technical writer position opened. I assume it started with the Egyptians or maybe the Atlantians. Someone had to describe the great deeds of the rulers; so regardless of the title, a technical writer position was created.

I pity the poor writer that had to chisel a noble's deeds into stone. The writer could not afford mistakes; after all, hieroglyphs were not easy to chisel and one wrong hammer blow could not only change what was supposed to be recorded for posterity but also destroy the stela or tablet. I can only assume that if the writer made a mistake, the head writer may have had the poor writer's head—literally.

When I first became interested in technical writing, a typewriter was the primary way to place printed words on paper. Text was typed and then paragraphs were glued onto art boards. If a graphic was needed, it was glued in the proper place with the text placed around the graphic and then the art boards were photographed for printing.

If a mistake was discovered, the text had to be retyped or Wite-Out was used to "erase" the offending word or words. Usually it was just easier to retype the paragraph and glue it in place. In those good old days, mistakes were costly both in time and in effort.

Realizing my typing was not of the highest quality, I searched for a better and more forgiving solution for creating user documentation. My search led me to the Apple Macintosh computer, which hit the market in January of 1984.

The Macintosh came with 128K of memory and a floppy disk drive. But what made the Mac useful for technical writing was its ability to combine text and graphics in a word processor document. The personal computer from IBM was not yet on the market, and when it did hit the market, it only ran on a primitive MS-DOS operating system.

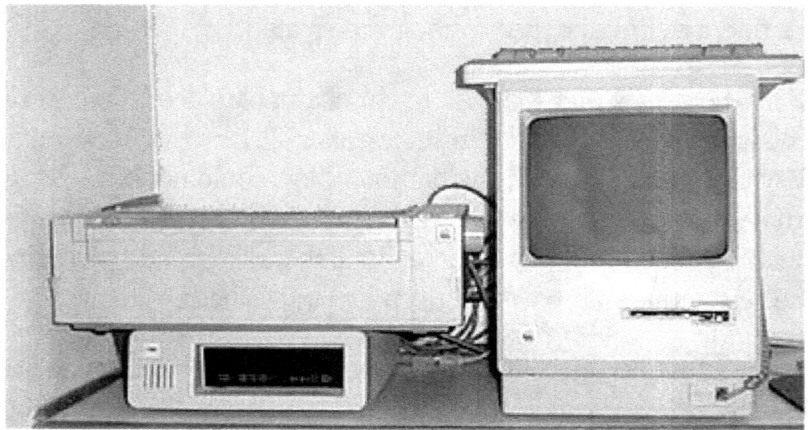

Figure 1–1: Authors Original Mac - Still Works

I bought a Mac and started my career as a freelance technical writer. With the Mac, a graphic could be created in MacPaint and then pasted into MacWrite. The days of cut and paste artwork were numbered and desktop publishing was born.

A document could be printed, using the ImageWriter printer that was part of the Mac's package. While the dot-matrix printer's output was not typewriter quality, it was neat and readable but best of all, changes could be made and the user documentation quickly reprinted. When Apple introduced the LaserWriter printer, output quality took an enormous jump. Copy approaching the quality of a printing press ensured that desktop publishing was here to stay.

User documentation that once took months to write, paste-up, and photograph for printing, could be written and printed in less than one-half the time and with little quality differences. This time difference was the main reason I got my first freelance technical writing job.

Writing user documentation is a form of self-publishing, at least it is today. Just as it was thousands of years ago, the writer was also the publisher. The manuscript was written on paper, animal skins, or even clay tablets and then dispersed for someone to read. The invention of the printing press, whether by the Chinese or by Gutenberg reduced the author to just writing; the printer determined the look of the final product. The printer determined the formatting, the font, and just about everything else except the words.

Publishing has come full circle. A technical writer is usually the publisher. The writer determines the formatting, selects the fonts, and decides how the user documentation looks, regardless of the publishing medium. While the writer may have to follow company guidelines that specify which font to use where or how to format the document, in most cases that writer controls the actual look of the published product.

What Is Technical Writing?

Technical writing is describing any subject, technical or not, so it can be understood by someone who has little or no training or knowledge of that subject. It is usually a formal or semi-formal style of writing. That is, the writer uses formal language, no slang or jargon unless that jargon is specific to the product or activity being described.

Technical writing generally refers to technological devices such as test equipment, but it can also include such household items as VCRs, TVs, microwave ovens, and refrigerators. Technical writing can also include anything from medical imaging products to computer operating guides to semiconductor test equipment.

Depending on the writer's skill level and training, a technical writer may be asked to write application help or content for a company's website and possibly even to assist in the design of that website.

Technical writing is not creative writing, although, at times it may seem that it is. New products can test a writer's creativity when documentation is required before the product actually works.

On the whole, technical writing is easier than fiction writing because the product directs your attention and narrows your scope. Fiction writing, on the other hand, requires more imagination because you start with an idea and use your imagination to turn that idea into reality for the reader.

However, this does not mean that a technical writer is any less creative or that the work is any less challenging, especially if required to write documents on products that are primarily concepts or as some say, "vaporware".

Author's Aside

The best learning experience is in small organizations where the writer may be the only one on staff. While this can be more challenging and difficult, small organizations usually allow one writer to completely write a user document. Also, the writer has a much better chance of writing many different types of documentation, from web content to marketing brochures to training materials.

A technical writer often toils in anonymity and as just one writer in a large group of writers, usually without seeing the finished product. In large organizations, technical writers are often assigned only a small portion of the final documentation. The head writer is responsible for pulling the various pieces together and editing them to ensure consistency throughout the company's documentation.

Most users learn what stirs their interest or what they need to become part of a group. Therefore, the task of a technical writer is to produce user documentation that is not only interesting but also instructive and informative.

This challenge is not as difficult for a writer in a large company that has the budget for buying the latest presentation and instructional products. However for a small company, getting the most out of what you have available in the least time is important and critical and can be a real challenge.

Technical Writer Requirements

A technical writer may have a degree in English or Journalism, but the best writers usually are those people who have experience in the field in which they write. It is easier to teach a technically trained person to write (even with some editorial help) than it is to teach a writer to understand the technology he is to write about. Writers with good technical knowledge are difficult to find and, therefore, can usually demand a higher wage.

Can I Be A Technical Writer?

Today, degrees often mean less because employers have discovered that a degree does not necessarily mean the person has the skill or the desire to do the job. If you are a good technician, with a little effort and work, you could become a technical writer.

However, before jumping into this new career ask yourself a few of the following questions. Also, make your own list based on your current position before deciding.

Here are few questions to ask yourself (be honest!):

1. Do I enjoy writing?
2. Can I sit at a desk most of the day?
3. Can I type? (Hunt and peck method is slow but works.)
4. Can I view the product as a customer or user?
5. Do I need to take writing classes?
6. Do I like annoying people?
7. Would I mind interviewing programmers and engineers I now work with to gather information?
8. Can my present employer use me, or will I have to look for a position elsewhere?

Before making a final, and possibly life-changing decision, talk to a technical writer and ask what he does and maybe even watch the writing progress. You may decide that what you are doing is much more challenging and interesting, or you may find a way to combine your current position and technical writing into a new and unique position.

> **Author's Aside**
>
> If you are serious about becoming a technical writer, ask your employer for an opportunity. Indicate a willingness to start a writing project, even at less pay, just to prove you can do it. If your employer likes your work record, your willingness to prove your ability will be appreciated.
>
> You may find you do not have the patience or aptitude for technical writing, or you may land a new exciting and challenging position.
>
> **Be willing to risk something to gain what you want.**

Technical Writer's Do's And Don'ts

A technical writer is often more than a just a writer. A technical writer should focus on the entire project, not just the documentation. Design engineers and programmers tend to focus on their specific duties; a technical writer must make an effort to see the complete project from a user's point if view.

As in any position, a technical writer is usually a person that wears many hats during the course of writing user documentation. Because a writer's duties and tasks change, the writer must be careful to do some things while avoiding others.

Some Do's

A technical writer often acts as an unofficial system analyst and verifies that all parts of the project fit together from a user's and customer's viewpoint. Often a writer can detect interface and operational problems and suggest solutions before these fixes become too impractical to correct or too costly to implement.

Designers and programmers are busy concentrating on specific issues and details; a technical writer concentrates on the overall project and those details that often are overlooked, especially in time-sensitive projects. A technical writer must view the product from the intended audience point of view. If written for the inexperienced user, the experienced user can skip, but the reverse is not true.

Attend all design review meetings and any brainstorming meetings—the more the writer knows about the product, the better he is equipped to pass on that information to an end user.

A technical writer's job is to be nosy and sometimes annoying. Getting information from busy engineers and programmers requires determination and stubbornness. People do not like to be bothered, but part of a writer's job is to extract information from busy people.

In a small company, this can be more entertaining because the writer has more freedom to float between departments. It is surprising what you can learn by watching and listening to conversations while problems are hashed out and solutions are worked through.

Be aware that most users prefer to talk to a person to resolve problems or to have questions answered. Regardless of how a document is made available or how well it is written or presented, most users prefer to call rather than to search for an answer. While electronic online media make the search easier, most users prefer talking to reading.

Therefore, the writer should consider if it is cost effective to write a user document. If programmers and engineers are receiving calls and e-mails from customers and users, a document might decrease the number of these calls and e-mails—saving money. Discuss the document's potential cost and its effectiveness in reducing these possible interruptions before choosing the manual's type and size.

NOTE: A user document is any form of product information that is intended for a customer or user.

Some Don'ts

A good writer can turn a 10-page manual into 20 or 30 pages, but is that a good idea? Increasing a manual's size can be done using several cheap tricks: increasing font size, increasing line spacing, increasing the space between paragraphs, decreasing the margins, and increasing the size of the graphics.

Adding unnecessary procedures is another method. One example would be to write a procedure for removing and replacing each board in a product instead of writing one procedure that works for all similar boards. Another method is to write as if you are paid by the word. (I have been *falsely* accused of this a few times.)

Most of us have read fiction books where the author writes several paragraphs about a blade of grass or about a sandy beach. This is almost always fill-writing: writing done to bloat the page and word count. This type writing quickly loses the reader's interest. What is true for fiction is also true for technical manuals and for all forms of user documentation.

Brevity is good but the intended audience must understand the document. If too brief, brevity is bad and could have disastrous results. There must be enough information so the user understands what to do; otherwise, he may decide to fill in the missing information as he sees fit. Finding the balance between too little and too much information is not easy, nor will everyone agree on which side the "too" lands.

Author's Aside

The first rule of writing, any writing, is to communicate. This is especially true for technical writers. If the intended audience does not understand the material, it is poor writing.

As a client once told me, "Our manual was written by an English professor, but we don't understand what he wrote. The manual must be rewritten."

Chapter 2

User Documentation Overview

Documentation Thoughts

Many companies are eliminating hardcopy manuals in favor of electronic variations: eBooks, PDFs, online HTML, application HTML help, videos, and other methods that avoid the expense of printing and shipping.

While I understand their reasoning, I believe they may be placing cost over convenience, or perhaps I am just old fashion and prefer the feel of paper to a monitor screen. Besides, dropping a paper manual does not harm the manual, but dropping a computer or an eBook reader can be a shattering experience.

HTML help is the major replacement for hardcopy manuals. Many companies, especially software only products, include only help with their products. Because of this and to reduce the cost of producing a manual plus application help, most help authoring tools provide for creating a manual from the help. (More in Chapter 13.)

Regardless of the final format of a manual or any user document, it should be treated as if it will be printed some day by some one. The complete document and specific pages should be easy to print and the user allowed to print any part of it he wants. Usually, a user prints those procedures that must be performed when the product is turned off.

TIP: Most help systems print only one topic at a time. This may be annoying to a user if several topics are needed for off-line repairs and adjustments. Including a PDF manual with the help and callable from the product's Help menu is helpful.

> **Author's Aside**
> There is research suggesting that reading a paper document is 25 to 30 percent faster than an electronic document, and reader comprehension is improved when reading a paper document.

What Is User Documentation?

User Documentation is any document written to help an end user understand and correctly use a product. The list in Table 2-1 is but a small sample of possible user documentation. Document types and styles vary based on either the company's standards or what their customers want or complain about.

Table 2-1: Examples Of User Documents	
• Interface guides	• Product requirements
• Assembly instructions	• Specifications
• Getting Started guides	• Training materials
• Maintenance procedures	• Troubleshooting guides
• Installation guides	• Tutorials
• References	• Application help
• Marketing Brochures	• Videos
• Presentations	• How To guides

Purpose Of User Documentation?

User documentation introduces a company to a prospective customer, defines what the company does, and reminds customers of the company's existence when marketing personnel are no longer around. Yes, it also tells and shows users how to operate a product.

User documentation has four primary reasons for being created:
- Educate the user,
- Decrease support costs,
- Market the product,
- Customer demands it.

If a document does not meet expectations in at least two of these four reasons, it may be a waste of the company's money or of the your time.

Importance Of User Documentation

User documentation is often the first product a potential customer sees. What that document says about the product and your company may help a customer decide if they should buy that product.

TIP: *User Documentation is a product!*

User documentation should:
- Be clearly and concisely written.
- Be easily understood by readers of average reading skills.
- Be written by writers who understand the products and can communicate that understanding to less experienced users.
- Tell users what to do if the product should malfunction.
- Suggest the consistency, effort, and quality put into the company's other products.

There is more to writing user documentation than just imparting information:

- Who is going to read that document: engineers, users, management, marketing, or secretaries?
- Is the document formatted so someone wants to read it?
- Is the document arranged so a reader can quickly find the information he wants?
- Are the company's user documents consistent from product to product?

Writing is as much an art as a technique, much like management, marketing, product design, and other disciplines that make a company successful.

Equipment Needed

Producing user documentation requires both hardware and software for the best results. Here are some items a technical writer needs:

Computer

Most businesses use the PC but the Mac should not be ruled out. PC hardware is usually less expensive. The Mac still appears to be more stable and less susceptible to viruses. However, the Mac interface is not the same as the PC's, so your decision may be based on which is easier for you to use or what your company dictates.

A monitor that allows the writer to view and read a full page at 100 percent is an advantage, but it is important to have a monitor large enough to view most of a standard page. Also, the pixel size is important when working with graphics.

If you require graphics, you may need a scanner.

Author's Aside
Install a virtual operating system that lets Mac programs run on a PC or PC programs to run on a Mac. (VMware and Microsoft Virtual PC are two programs for the PC. Boot Camp and Parallels run on the Mac.) If installing a virtual OS, have someone do it that has experience. These programs can be tricky and can mess up your computer if improperly installed. Make a backup of your computer's hard drive before installing a virtual OS.

Software
Word Processor: Word and WordPerfect are two major word processors. The word processor may be a company decision.

TIP: A word processor should include a spell checker, a grammar checker, and a thesaurus.

Do not start with Pagemaker or QuarkXPress because they are primarily intended for publishing, not for authoring. FrameMaker can be used for both authoring and publishing, but it is expensive compared to just a word processor and it is more difficult to learn. However, it is great for large manuals with complex formatting.

TIP: Install a quality anti-virus program and a firewall or router to protect your files and your computer from hackers.

Graphics Program: Creating user documents requires working with graphics such as screen captures, charts, and graphs. While many programs can do this, Photoshop still appears to be one of the best all around programs for reworking, sizing, and converting from one graphic extension to another.

NOTE: Photoshop may not be the best tool for all applications and unless working with a lot of pictures or screen captures, a less expensive tool may be a better choice.

TIP: Avoid using BMPs as they tend to be large in file size, do not size well, and do not display or print as well as JPEGs. (Resolution should be 200 DPI or better for quality printing. 72 to 150 works well for the web and monitor viewing.)

Adobe Acrobat: There are online sources for converting a file to a PDF, but it is better to have Acrobat installed on your computer, not just the Reader. An older version may work but your company may insist on a newer version.

Help and e-learning tools: there are many programs that can be used in this category, depending on your budget and your output requirement. I am familiar with Fast-Help for creating HTML help and WinHelp files. Adobe's Captivate and Techsmith's Camtasia are two good screen recorders.

Screen Capture: A good screen capture program can save a lot of time; however, Windows **Print Screen** and the **Alt+Print Screen** work quite well. Use a good graphic program to proportionally scale and convert to a JPG. You can also paste directly into Photoshop, convert the image and scale it for use.

NOTE: Screen captures are usually 72 to 100 DPI. A good graphic program can increase the DPI to 200 or better. Avoid scaling a BMP in a word processor. Use JPEGs instead.

Recording Device

Because your memory is not perfect and unless you are great at shorthand, buy a good recording device. A tape recorder is less expensive but more difficult for you to find the exact data you want when you want it. A digital recorder allows you to create bookmarks and also allows you to record in different electronic folders, making data organization and retrieval easier and faster.

TIP: Some digital recorders are capable of downloading to a computer and converting it to on-screen text.

Note Pads

Yes, old fashion note pads. Note taking is still a necessary requirement. Use one note pad for each product to write ideas as you review the product, perhaps even as you write parts of the manual.

TIP: A 6 by 9 Steno Book type is easy to carry.

Upgrade Carefully

There is a tendency for many writers to want the latest and greatest in computer hardware and software, but there can be problems in pursuing the latest updates.

Occasionally, installing an operating system's updates or service packs can cause some programs to quit working or to become unstable. Be careful going from one operating system to another. Applications that work under Windows 95 or even XP may not work under Vista unless the latest versions of those applications are bought. This is also true for the Mac.

Peripherals not working with a new operation system are a common problem. The new operating system may not come with the drivers required to use that favorite printer and scanner and their manufacturers may not have new drivers available—yet.

Do not upgrade programs simply because newer versions are available unless the upgrades provide features or functions that can improve your performance and decrease your workload.

Sometimes a program's new version departs significantly from its previous version and learning the new version is the same as learning a new product. If a product or application works for you, do not discard it simply because a newer version is being pushed. The older version may be more intuitive to you or some features may have been relocated, making them more difficult to find.

Evaluate then decide.

Chapter 3

Manual Types

Manual Types Overview

Manual types are changing so that one manual serves as both an Operator's Guide and as a User Manual, as one example. This chapter lists and describes several of the most requested manuals. Also included in this chapter is a description of how a typical manual is organized.

Manual Types

A list of the most common product manuals:

> User Manuals or Operator's Guides
>
> Getting Started Manuals
>
> References Manuals
>
> Theory Manuals
>
> Troubleshooting Manuals

User Manuals or Operator's Guides are written for inexperienced users. They describe all controls, screens, and basic functions that an operator needs to operate the product. These manuals should include system requirements such as special installation requirements unless a separate Installation Manual is provided. These manual should include only those procedures an operator is expected to perform.

Getting Started Manuals are written as a "How do I get the product working the way it is supposed to as quickly as possible?" manual. These manuals should include initial setup information as well as a simple overview of the product. If parameters must be input before the product is operated, that information must be defined. These manuals contain less information than a User Manual or an Operator's Guide, but they should describe those controls and screens necessary to prove the product is operating correctly. They may also include how to install the product.

Reference Manuals are written for experienced users and perhaps programmers if code is involved. They describe those functions that an inexperienced user is not expected to perform or understand. They contain in-depth details such as code descriptions, error messages, and procedures that are rarely performed. This manual is primarily needed for software products and proprietary user interfaces.

Theory Manuals are written for technicians and engineers responsible for replacing components and repairing the product. They often contain circuit schematics and a description of each circuit. They may also include troubleshooting information.

TIP: If schematics are included in a theory manual or are in a separate manual, use 11 by 17 pages or show only small portions of large schematics on standard or smaller size pages.

Be sure to include the version number of the schematic. The written theory and the schematic seldom match unless the theory is updated when the schematic is updated.

Troubleshooting Manuals are written for technicians and engineers. These manuals are usually combined with theory manuals and contain charts that indicate if this problem exists, look at these possibilities. They are often used to find the problem and what to do to fix that problem. These manuals are normally used to isolate problems to the board level, while theory manuals are used to isolate problems to a board's individual circuit.

NOTE: It is common for customers to isolate product problems to a board and to replace that board while the failed board is returned to the manufacturer for repair. Production downtime MUST be minimized.

This is only a partial list of possible manuals. You may wish to merge one or more of these into one manual, depending on the information needed. You may also decide multiple, small and very specific manuals are the better solution.

Author's Aside

I have found merging manuals is usually easier than creating several small manuals. The more manuals there are, the more chances they will be misplaced even if they are in electronic formats. Also, having to find several manuals to install and setup a product is annoying.

Typical Manual Organization

Some technical writers suggest organizing manuals so simple and frequently performed tasks are placed before more complex tasks. The problem with this approach is that what is simple or complex depends on the user, not the product or the writer.

Organizing a manual based on simple procedures and frequent tasks may depart from how the product's interface is organized. Ordering tasks this way may work for some products, but this type organization may confuse the user who must become familiar with the product's interface.

As an example, a frequent task may be copying database files to the network or to an external hard drive, but that cannot occur until tests are run and data is available to be copied. So, why show the user how to do this task before the user is shown how to perform a test or how to enter test specifications?

Most technical manuals follow the same organizational pattern of a typical textbook. Technical manuals, as do most books, have three main divisions:

- Front Matter
- Text Block
- End Matter

Suggested materials for each division are included in the following chapters. The content of each division varies with the manual type, but the material is usually predictable.

Chapter 4

Front Matter Material

Front Matter Overview

This chapter describes a typical technical manual's Front Matter material. Included are formatting suggestions as well as content suggestions. The actual content varies with the manual type, especially the Preface.

Most Front Matter materials, except the Table of Contents, are boilerplate. That is, once written, the material can be copied to another manual of a similar type with only minor modifications.

Front Cover

Because a manual's front cover is often the first thing a user sees, it provides the first impression of your company and its products. This is especially true for printed manuals, but electronic manual covers are also important.

Design the cover to look good in black and white with gray tones. If the cover is printed in color, it will look that much better. Using color on the cover is always a good idea. Color attracts the eye and can generate interest if properly used.

TIP: Avoid red with black text as both appear black when printed in black and white. Test color combinations before deciding.

While a technical manual cover cannot be compared to a fiction or textbook cover, the principles are the same: If the cover is boring, readers may assume the rest of the book is boring. The cover should not be used to sell the product, only to identify it.

Unlike a fiction book, a technical manual is not usually read for enjoyment, but that is no reason to neglect the cover's design. The budget allowed for the manual often determines the time and effort available for cover design. This is one reason why similar products should have similar covers.

Outside Front Cover

Figure 4–1: Minimum Front Cover Information

Company or Product Logo: The company may specify the location of this item, but it should be prominently displayed. A picture of the product could be added or could replace the logo.

Product Model Number: This is the control number of the product. Some companies may not want the model number displayed and may only want the product's name and description.

Product Title or Description: If the product has a descriptive name or some other unique description, put it here.

Company Name & Address: Place the company's name and address on the front cover where it can be quickly found. Some companies may prefer this information be placed on the back cover or in a specific customer service area of the manual.

Inside Front Cover

Figure 4–2: Typical Back of Front Cover Information

TIP: Including a header and footer on the back of the front cover is optional but advisable.

Product Warranty: Some products may have only a hardware or a software warranty, but including both is best. Making the back of the front cover generic saves time, especially if there are other similar products. Warranty statements must be provided by the company and cannot be changed without permission.

Copyright Notice: Use the form: "Copyright © year by Company Name, City, State". Add either a statement similar to the following or one provided by the company: "Printed in the United States of America. All rights reserved. Do not reproduce any contents of this publication in whole or in part in any form, by any means without the prior written permission of 'Company Name' ".

Trademarks: List the product the manual is written for as well as any products that are mentioned in the manual. This includes the company's trademarked items and products.

Examples:
IBM is a registered trademark of International Business Machines Corporation.
MS-DOS is a registered trademark of Microsoft Corporation.

NOTE: There is a trend to not mention trademarks except to indicate they are correctly identified in the manual.

Manual Name & Part Number: This is for manual tracking and identifies the manual by name and company part number.

Other credits and disclaimers may be required based on the product and if the product contains a component that must be disposed in a specific manner. An example might be a component that contains a toxic or hazardous material; however, this type information should also be placed in the Preface.

Obviously, the exact look and design of the manual's front cover changes based on the manual's page size and the number of products expected to be sold. If the product is a household item, the manual may be professionally printed and bound. If the product is not expected to sell more than 20 to 50 units, the manual may be printed in-house or sent only as an electronic file.

Reader's Comment Page

This Front Matter page may be included in low volume, highly complex products. A Reader's Comment page allows the user or customer to send suggestions about the manual to the company's documentation department.

Include the company's mailing address and the department to which this page should be referred. If the manual is intended to be only an electronic document, consider including an e-mail link so the user does not have to print the page.

This page lets the reader know that your company cares about how the manual works and appeals to its customers. It is seldom filled out and returned because the user usually complains to sales, customer service, or to training personnel.

Even if this page is never used, it is worth the added effort unless the company prefers it to be omitted.

Table Of Contents

A table of contents shows the topics in each chapter. The easiest way to construct a TOC is to use the word processor to assign Heading 1 through Heading 4. The word processor can then read those Headings and construct a table of contents. Word uses TOC 1 through TOC 9 that can be formatted to display the contents as you prefer. (Avoid more than four Heading levels.)

Headings Example 1:

The above example shows the formatting for a typical table of contents; however, for a technical manual it is better to provide additional information. One thing a user should know is which chapters should be read based on what that user is expected to do with the product.

As an example, a plant engineer needs to know what power source is required, if a clean room is needed, or if special plant facilities are required. An operator may not need to know these things but wants to know how to operate the product. A technician needs to know how to calibrate the product and how often.

Figure 4-3 show a suggested table of table of contents layout, using two-columns.

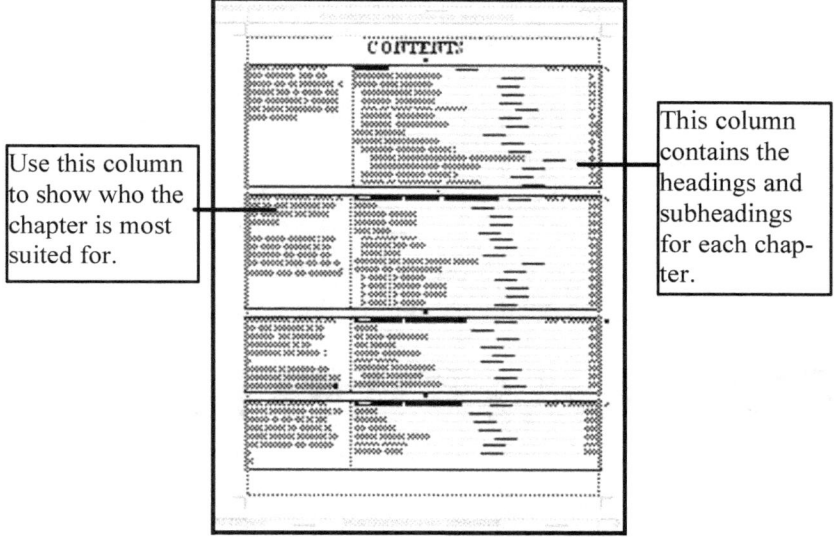

Use this column to show who the chapter is most suited for.

This column contains the headings and subheadings for each chapter.

Figure 4–3: Two-Column Contents Page

NOTE: The two-column layout is more work than the one shown on the previous page, but the two-column layout contains information valuable to the user.

Depending on the number of pages and importance of the Preface, the Table Of Contents may come before or after the Preface. If the Preface contains important information, its contents should be listed in the Table Of Contents. If the Preface is short with no safety information, it may precede the Table Of Contents.

There is a trend of placing a list of figures and tables in the Table Of Contents. Most users do not look or search for a figure or a table; instead, they look for the description or instruction that figure or table presents. Word processors have made adding these lists simple and some writers assume they create a professional look.

While there is some debate about figure and table lists, there is little debate about adding a list of "How To" procedures to the Table of Contents. However, a better solution is to provide a separate chapter or brochure devoted only to "How To" procedures. These same procedures should be in the manual where they show the user what steps to do as the topic is described, but by also grouping them in one chapter or in a separate brochure, the experienced user can quickly find the exact procedure needed.

TIP: If the "How To" procedures are grouped in a manual, they should be grouped the same way in the help system.

Preface

The significance of the Preface depends on the type manual being written. The Preface may be only a paragraph or two for a low page count manual and may not exist in a household appliance manual or brochure.

It is not unusual for a low page count manual to omit the Preface, but the Preface is important for test equipment manuals. In this type manual, the Preface may contain everything from power safety warnings, to earthquake safety information, to robotics safety information. The amount and type of information included in a Preface depends on the product, where it is being shipped, and the manufacturer's, and even the customer's safety policies.

The following is a list of possible information that may be required for inclusion in a technical manual's Preface.

Customer Assistance: Include support's address and telephone number for the user to contact in case of questions or problems.

About This Manual: Include a proprietary notice as a second level of protection for the manual.

Manual Overview: Describe what the manual is intended to do. Does it tell how to operate or how to repair the product and what type training the user needs to correctly operate the product?

How To Read The Manual: Suggest which chapters should be read first, by whom, and why.

Manual Revisions: Describe how revisions are made to the manual. Will the manual be updated as a unit or chapter-by-chapter? Also, where the revision code is located. (Large printed manuals are best updated chapter-by-chapter. Electronic manuals (PDFs) are usually updated as a unit.)

Manual Conventions: Describe the difference between **Caution**, **Warning**, **Danger**, **Attention**, and **Note** statements used in the manual. Show multiple keystrokes: **CTR+ALT+DEL** and other inputs the user may be required to perform. This paragraph is more important for hardware products than for software products.

User License: Ask marketing or the company lawyer for this. If the product uses a company written user interface or proprietary firmware, a User License is usually part of the sales contract.

Safety Instructions: This can include instructions for **Power Lockout/Tagout**, which is usually required to meet certain certifications. These instructions are needed in a production environment where one person may take the product out of operation and the product cannot be returned to operation without that person's knowledge and permission.

Describe any safeties or interlocks the product may have. This can include anything from a Kill Switch to a Tilt Switch.

Robotics Safety: This warning may be required if the product can operate without an attending operator. A warning is especially necessary if the product can cause harm to an inattentive operator. Graphics should show the restricted areas when the product is in operation or when power is applied.

General Safety Rules: These are rules that should be followed when the product is installed and connected to plant power. If there are voltages capable of causing injury or death, a warning should be placed here as well as on the product's case.

Types of Electrical Hazards: This information may be required to meet some safety and certification guidelines.

Emergency Turn Off Conditions: Emergency conditions should be described as well as how to restart and return the product to operation.

Earthquake Safety: These instructions may be required, depending on the size of the product, where it is being shipped, and perhaps on the customer. Diagrams that show how the product should be tied down and even the product's **Center of Gravity** may be needed if the product is of a certain height and weight.

NOTICE:
Some of the safety instructions listed for placement in the Preface may be placed in a special chapter of the manual, depending on company guidelines and preferences.

In low page count manuals, the power safety instructions are often placed on the back of the manual's front cover and other important safety instructions may be placed on the first page of the manual. This is usually the case if a Preface is not included.

Regardless of where these instructions are placed, they should be referenced anyplace that is appropriate in the manual.

While all of these safety instructions may not be needed or even necessary, one of the primary duties of a technical writer is to limit the company's legal liability that may result from inattentive or inexperienced operators and users.

NOTE: Even if a manual is not written as a separate entity, Front Matter should be included. If the manual is created from the help, Front Matter can be added either to the help or added only for a printed manual. (See Chapter 13)

This Page Intentionally Blank

Chapter 5

Text Block Material

Text Block Overview

The Text Block is the heart of any manual because it describes what the product does and how to operate it correctly. This section of the manual may be a few paragraphs to 10 or more chapters, depending on the product's complexity.

One of the first problems facing a writer is how to organize the manual so a user can quickly find what is needed for a particular task. For this reason, consistency from manual to manual for similar products is a priority. This means if Installation is Chapter 3 in one manual, avoid placing that same information in Chapter 10 or in an Appendix in another similar product manual.

You should have decided if one or more manuals are required or if only a help system is needed. If the product is extremely complex and performs several diverse functions, it may be better to create manuals specific to those functions instead of trying to cram all of the material into one large, ungainly manual.

TIP: Thin manuals are less intimidating.

Hardware Or Software?

Next, determine if the product is primarily hardware or software. The reason for this decision is that the product often determines if the manual or the help is the first document started.

The easiest way to decide if a product is hardware or software or a combination of both is if the software operates independently from the product's hardware. Why the distinction between hardware and software? It is usually easier and better to create a manual first for a combination hardware-software product even if it runs some type of primitive help system. If the product is primarily software, it may be faster and easier to create the help and then use the help to generate the manual.

To help decide, answer a few questions about the product:

1. Does the hardware have to be described?
2. Can the product display WinHelp or HTML Help?
3. Can the product display an online manual: PDF or other?
4. Can the software operate without the hardware?

Does the hardware have to be described in order to describe how to operate the product or do you only need to explain the software's user interface? If the hardware must be explained, start with the manual.

If the product can run either WinHelp or HTML Help, the product could be treated as either hardware or software, and it may be better to create the help and then generate a manual from that help. (This assumes the software is mostly complete.)

If the product cannot support a normal help system but can run a primitive help system, the product should be treated as hardware and writing the manual first may make creating the help easier.

Also, if the software depends on the hardware to operate, the product should be treated primarily as hardware and a manual written first. And, of course, if the product is only software, help may be the only viable option.

> **Author's Aside**
> The decision of whether to start the manual or the help may depend on the documentation's deadline and if the software is at the stage where it can be run so help can be written.
>
> Help should not be written until the software is at that stage that major changes are unnecessary. However, some boilerplate parts of help can be written before the software is at that stage. A few possibilities are: Welcome To..., Introducing..., and What's New.

Hardware Text Block

The following are some suggestions as to the content of the first chapters of a hardware product.

Chapter 1: Introduction

Introduction or a similar chapter should be the first chapter after the Front Matter. This chapter should provide an overview of the product, what it is used for, and other details that describe the product without sounding like a sales brochure.

This chapter should provide a brief but accurate description of the product's operation, how a customer should use it, and what the product does that is useful to the customer.

If the product performs a test that is defined by a recognizable organization, such as a Mil-Spec or a JEDEC Standard, identify that test specification or standard.

Include a picture of the product. Create a list of its features and how it differs from a previous model, if one exists. Provide a short description of each test the product performs.

Write this introductory chapter for busy engineers and managers who need to know if the product does what they need it to do but not exactly how it works. In other words, brief and vague but enough to indicate that the product can test the components the customer wants to test.

TIP: This chapter is not for describing specific operations.

Do not define any front panel display and switches in this chapter. Define these items in detail in a **Getting Started** or a **Basic Operation** type chapter (Chapter 4).

Chapter 2: Specifications

The second chapter provides the specifications of the product. It should list any built-in product safety features that protect the operator or the equipment. An example might be that two pushbuttons have to be pressed at the same time to start a test.

Another example may be that if power is interrupted, the product's On/Off front panel switch has to be turned off then back on to restore power to the unit.

Consider the following list of possible specifications for inclusion in this chapter.

Table 5–1: Product Specifications	
Power Requirements	120 or 220 Vac, single-phase, three-phase, 50 or 60 Hz, special connector
Product Clearance	Show distance from product to wall, room required for servicing, using a diagram
Facilities Requirements	Min/Max PSI for compressed air, Inches of Mercury for vacuum, cooling ducting
Dimensions	Height, Width, Depth, and Weight
Environmental Requirements	Operating temperature range, Storage temperature range, Maximum humidity for operating and storage
Tests Performed	A brief list of the tests performed by the product even though an expanded description may have been included in the Introduction chapter
Test Modes	Manual, Automatic, Production, Engineering, etc.

Chapter 3: Installation

The third chapter describes how to install the product. Installation may be as simple as applying power, or it may be as complex as connecting several cables and additional test equipment such as a DVM (Digital Volt Meter) or an oscilloscope.

Here are some suggestions for a checklist:

- Inspect packages for shipping damage.
- Notify carrier if damage occurred during shipping.
- Check each package against packing list.
- Provide contact information to report problems.
- Suggest best site for placing the product.

- Level of experience required for installing the product.
  - Can customer's technician install?
  - Should only factory-trained personnel install?

Next, where and what connections must be made to the product. This is best done using diagrams or pictures of the connection panels. It may be necessary to show pictures of the cables and indicate the exact mated ends and connectors. If the designer has done a good labeling job, each connector is identified and each cable has an identification number; if not, suggest these identifiers be added to each cable and connector.

TIP: If connectors are of the same shape and size, color coding them helps the user in correctly mating the connections.

If the product consists of several pieces of equipment that must be connected together, provide a diagram that clearly shows how the individual units must be connected and how each unit should be located in reference to the other units.

Describe the sequence in which the product is power up. Is it important to turn on one piece of equipment before another?

Does any software or firmware need to be updated or installed before operation? And finally, indicate if calibration must be done before the product is used or if there is a suggested warm-up period.

Chapter 4: Getting Started
This fourth chapter should be **Getting Started** or a **Basic Operation** type chapter. This is where the detailed description of the product's operation begins. One of the first things to describe is how to apply power even though that may have been described in a previous Installation chapter. By repeating the information here,

the user does not have to look for it elsewhere, even though applying power may be as simple as flipping a power switch.

TIP: It is better to duplicate information than make the user search for it or guess at how something is done.

If the product has front panel controls, their operation and function are described. This is best done using a two-column table that lists the control and then what each control does.

If the product has a display, describe its operation, what it displays, and any information that protects that display. As an example, if the product uses a touch screen, its operation should be described, and a warning about using sharp objects to activate a screen selection should also be included. (Remember, people who are not familiar with that product's technology may operate the product.)

TIP: Touch screens are very popular because a keyboard is not required. Smartphones are but one example.

Use a simplified block diagram of the product's test circuit to describe how the product performs its specified test or tests. The test description should be detailed enough to show the required certified standards are being met but not so detailed as to give away proprietary circuit design information.

A flowchart of a test cycle provides a user with a sense of the steps being performed. Also, depending on the product's functions, test waveforms may be helpful in describing exactly how the product works or how the product decides if a component passes or fails.

It is strongly suggested that a **Getting Started** list be provided that walks a user through the exact sequence needed to place the product into operation. This means that if certain parameters must be entered before the product can be used as intended, those screens and the chapters describing those parameters should be listed and those descriptions placed in the appropriate chapters.

A Menu Interface description may be needed, especially for a proprietary interface that does not look or act like a standard Windows, Linux, or Mac interface. If needed, briefly describe what each menu does, what each icon calls, and any screen feature that a user may not understand. If a typical user knows how to operate a Windows interface (as an example), a menu interface may not be required except for non-standard menu items.

NOTE: If the menu interface is complex and has many functions, it may be better to add a separate Menu Interface description chapter.

When the product is turned on, briefly describe any screens that display before the main menu is displayed. A pre-screen example could be a self-test screen that shows if the product is operational or if there is a problem such as a low voltage. Do the same for the Main Menu screen and for all other major screens.

If help is available for the product's screens, describe how help is activated and how to use it. Describe any special icons or anything that might be unusual to an inexperienced user.

After these preliminary chapters are written, the next step is to decide the organization of the remaining chapters. As an aid in creating a user type manual, two combination hardware-software examples are provided. These examples have unique interfaces and have several main menus with several sub-menus. Use the main menu as much as possible to determine the order of the remaining manual chapters.

TIP: A manual should not express the writer's creativity; it should help the reader without calling undue attention to itself.

ITC55100 UIS Tester

The tester shown in Figure 5-1 needs a manual, whether it is a paper or an online manual does not matter at this stage. Because the product uses a low level operating system, it is limited to screen help and it cannot use either WinHelp or HTML Help. It has a very simple web-type HTML help that is called from each screen.

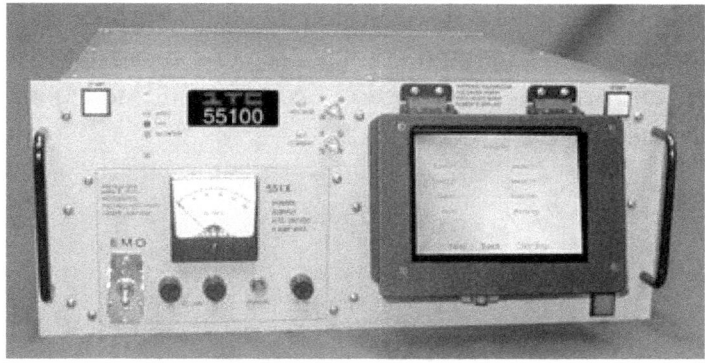

Figure 5–1: ITC55100 Tester

While this type help is not as sophisticated as a good help system, it provides the user with screen-specific information. Also, the manual cannot be included within the product's coding because of limited memory and display capabilities.

It is suggested that the chapters of this manual be named identical to the menu selections so the user can quickly locate a chapter that describes a specific menu function.

NOTE: Integrated Technology Corporation of Tempe, Arizona, USA has granted permission to use two of its semiconductor testers, ITC55100 and ITC59000, as examples.

The ITC55100 is a semiconductor tester that does not require a user to log in, which would be one of the first things described if it were available.

If the programmer and the product designer have done a good job of organizing the main screen, the menu selections are close to the order a user should follow in setting up test parameters and for testing devices.

The first menu item is **Test Specifications** and it should be the first item described and the next chapter after **Chapter 4.** Unless there are several large and complex screens called by the **Test Specifications** menu selection, combine all functions and screens called by it into one chapter.

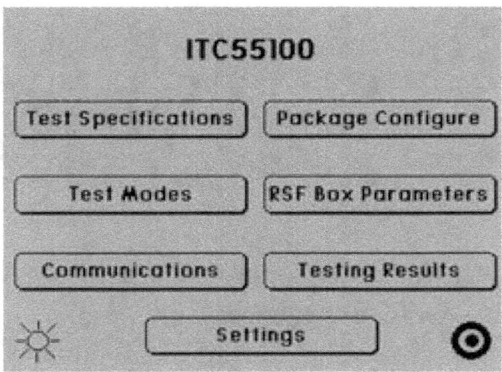

Figure 5–2: ITC55100 Main Menu Screen

NOTE: Test Specifications are those values and parameters that must be input so the product (tester) can correctly test a device.

The functions under the **Test Specifications** menu should be described in this chapter in the same order they appear on that sub-menu's screen. Again, this allows the user to quickly understand that the manual's descriptions and the product's menus are identical in name and order.

While this concept is simple, it is not uncommon for a technical writer to rearrange the manual's order based on the assumption of what the user needs to know and when the user needs to know that information.

This "writer-knows-best" attitude sometimes works for extremely complex products that have multiple sub-menu screens called from one Main Menu selection. When confronted with a number of sub-menu screens, it may be best to divide the manual into functional descriptions. This is normal for software only products and complex combination hardware-software products.

Also, if each of the Main Menu selections call several sub-screens and each sub-screen requires detailed information or the inputting of parameters to those screens, separate manuals that describe these functions might be a better approach but only if those functions are separate in operation.

For this example, let's assume one chapter for each Main Menu selection is sufficient. This means the following chapters are required. (See Figure 5-2, previous page)

> Chapter 6 = Package Configuration
> Chapter 7 = Test Modes
> Chapter 8 = RSF Box Parameters
> Chapter 9 = Test Results
> Chapter 10 = Communication and Setting

Notice that Chapters 9 and 10 are out of order with the menu structure. This is a decision based on keeping **Test Results** close to the tester's primary functions.

Chapter 10 is composed of two menu selections on the assumption that both of these selections require only a couple of paragraphs to describe them. If that assumption proves incorrect, these menu items can be divided into separate chapters.

A list in Chapter 4 **(Getting Started)** should have indicated the steps and the order needed to place the product in operation.

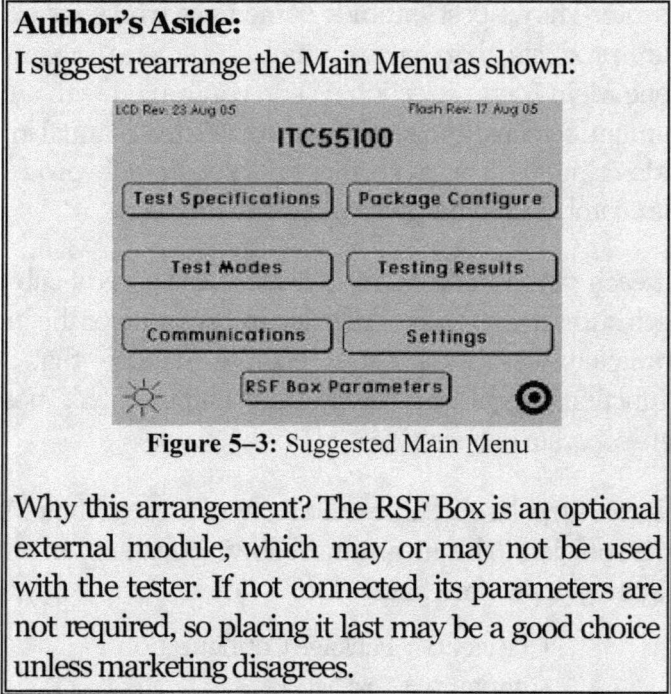

Author's Aside:
I suggest rearrange the Main Menu as shown:

Figure 5–3: Suggested Main Menu

Why this arrangement? The RSF Box is an optional external module, which may or may not be used with the tester. If not connected, its parameters are not required, so placing it last may be a good choice unless marketing disagrees.

ITC59000 Test Platform System
A second and more complex manual organization example is for an ITC59000 tester. There are several parts to this tester. This tester should be considered as hardware even though the tester runs an HTML help system under a Windows operating system.

For this example, let us assume the interface and control software are not yet ready but some screens are considered done for now. Therefore, we will start the manual instead of the help even though some parts of the help could be written without needing to use any help calls from the tester's software.

To decide on the type and number of manuals, the tester must be looked at based on its individual functions or tests. First there is the chassis, which can contain a maximum of four test modules. The chassis's hardware and software act as a unit to control, monitor, and evaluate test results from the individual test modules. The chassis does not perform any tests and it cannot function alone, and the test modules are dependent on the chassis.

Second are the test modules. The four individual test modules can be of the same type or four different types. Each test module should be considered as a separate tester and, therefore, treated individually with the chassis thought as a host or mainframe.

The question "Is one manual better than multiple manuals?" can now be answered. Because there could be up to four types of plug-in modules and each module can perform different tests, it appears best to plan on one manual for the chassis and one manual for each type test module. This approach also allows for future test modules with only minor changes to the chassis manual and to the individual test module manuals.

The advantage to one manual is that a customer would receive a manual with all test modules described, which may encourage that customer to buy the other test modules. The disadvantage is that the manual has to be much larger and contain information that is useless and unnecessary to a user that only has one type test module installed.

TIP: Consider the advantages and disadvantages, and perhaps consult marketing before making a decision.

This tester allows for user login to control each user's access to the tester's functions. Therefore, the first thing is to define each user's permission level and what they are allowed to access. This means that Chapter 5 should be named **Creating Users, Working With Users** or some other name that describes how to enter users and assign their permission levels.

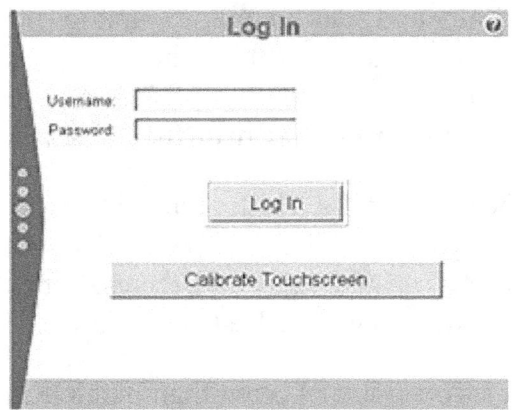

Figure 5–4: ITC59000 Log In Dialog Box

If the **Log In** button is touched (Figure 5-4) in the Log In box, the user is logged in with administrator privileges. Because the chassis has the controlling software, we will look only at the chassis **Startup** screen to see how to organize this type manual.

Notice that while the **Users** menu selection is last on the **Setup** Menu (See Figure 5-5, next page), it should be the first function described. If the customer decides not to create user permissions, this chapter can be skipped.

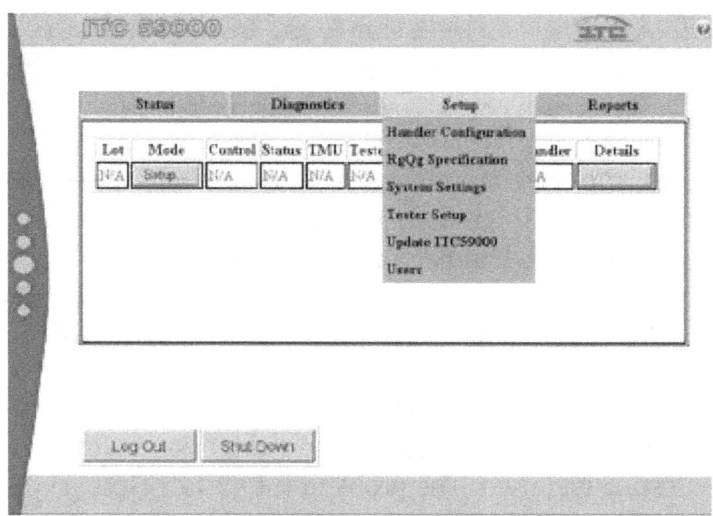

Figure 5–5: ITC59000 Start Up Screen

NOTE: The **Users** function is last because, once setup, it is accessed only by the administrator, not operators.

The steps that a user should follow to setup the tester for normal operation follows the menu structure by about 85 percent. For this reason, Chapter 4 should have included a **Getting Started** list so the user setting up the product knows which steps and in what sequence they should be performed for the most efficient setup.

Once **Users** are described, follow the **Setup** Menu to determine the next chapters in the manual:

> Chapter 6 = Handler Configuration
>
> Chapter 7 = RgQg Specification
>
> Chapter 8 = Creating Test Lots
>
> Chapter 9 = Reports

NOTE: A handler can be as simple as a test socket or as complex as an automated machine that takes parts from a bin and places each part into a test socket for testing.

For this example, let us assume that a handler must be defined before a test specification is defined. This means that when creating a test specification, a handler must be configured and associated with that test specification; therefore, the handler and its setup must be described before describing **RgQg Specification**.

Because the tester is intended to work in a production environment, it requires that devices to be tested be assigned to a test lot before testing is started. The problem is that **Lot Setup** is not on the setup menu. So before going further, the user needs to know how to create a test lot.

NOTE: Test Lots should not come as a surprise to the user if Chapter 4 **(Getting Started)** was read, which should have described all functions on the **Start Up** screen.

Creating Lots or **Working With Lots** should be Chapter 8 because once a handler is configured and a test specification is input, a device can be tested.

Chapter 9 should describe the **Reports** Menu functions because that is the natural order. A report cannot be generated until a device is tested and the test results are stored.

The remaining selections on the **Setup** Menu are primarily for an Administrator; therefore, they should be described in chapters following the **Reports** chapter. These administrator functions may not be available to most users if the customer has set up user permissions.

NOTE: Most of this manual organization can easily be applied to the help's organization.

Helpful Suggestions

Some suggestions when describing items on a screen:

- Describe what the item or function does, sets, or displays.
- Use the exact name as shown on the screen, even if the spelling is not accurate (suggest a fix) or is abbreviated.
- Indicate the ranges of allowable inputs—do not make the user guess. If the software protects against over- and under-range inputs, still supply the allowable range in the manual.
- Indicate the minimum step size and resolution.
- Does the screen have online help available?

NOTE: If online help is available, ensure the help agrees with the manual, especially if written by different writers.

Input Examples:

Test Voltage:	Sets the voltage at which the component is tested. (Is 1 volt entered as 1.0 or as 1000 mV?)
Range:	2.0-volts to 100.0-volts in 0.1-volt increments. Default = 1.1 volts; 0 = invalid

Date:	Sets the date displayed on the main screen. Stored in the test results. (12- or 24-hour input?)
Input Method:	dd/mm/yyyy (D = day, M = Month, Y = year)

Time:	Sets time displayed on the main screen and also stored in the test results.
Input Method:	hh/mm/ss for 12-hour clock time. (H = hour, M = minutes, S = seconds

NOTE: Date and Time must be correctly set in the tester before a test can be initiated.

> **Author's Aside:**
> Even when writing separate manuals that describe individual functions, it is still best to follow the menu structure as much as possible.
>
> If the menu structure is: **File, Edit, Format**, etc. and one of them is skipped, the user may want to know what that menu does if it was not mentioned.

NOTE: When the manual's Text Block is in its 2nd or 3rd draft, have the design engineer read it. This provides a check of your accuracy of the product's setup and operation.

Software Text Block

Software products have standardized on the main or top menu names and on most of the functions that appear under these menus. This has occurred because of the advent of the PC and the Mac. By using the same menu names, the user knows what to expect, regardless of the program's function.

Because of this standardization, top menus are not always a good indication of how to organize a manual or the help for a program. For this reason, the writer must first understand exactly what the primary purpose of the program is and describe that purpose first.

As one example, a word processor's primary purpose is to allow a user to create a document; therefore, the first thing to be presented is how to use the program to create a document. This means the manual and the help must first show the user how to open, name, and save a new document. Next, the user must be told how to format that document: fonts, styling, margins, and spacing.

Functions such as creating tables and using the drawing tools should be described later because those functions are things that add to the primary purpose of the program and are not essential to creating a very basic document.

Software Manual Sections

Because most software products deliver help as the only user document, the divisions of a software manual or help system are different from a hardware product's manual and help system. To show these differences a list of the subsections and their location is included.

Front Matter

Some Front Matter subsections are not absolutely necessary if help is the only user document, but it may be added to the help if a manual is generated from that help. Most help tools allow topics to be written that can be left out of the help but included in a manual output.

Subsections	Materials
Front Cover-Front:	This information should be included in a special help topic that is displayed when help is called from the top Help menu.
	Company Logo, Company Name and Address, Product Model Number, Product Name.
Front Cover-Back:	Include in a special help topic but may be placed at the bottom of the help divisions. Separate topics for each warranty.
	Hardware Warranty, Software Warranty and Disclaimer, Copyright Notice, Trademarks.

Subsections	Materials
Reader's Comment:	Should be a separate help topic at the bottom of the help divisions.
	Questions about the usefulness of the manual, suggestions for manual improvement, manual name, product model and name, and company mailing address.
Table Of Contents:	This is not needed as it is replaced by the help's books and topics Contents pane.
Preface:	This division is not usually needed for a software only product but is needed for a combination product. Create a separate help book and label each topic, especially the safety warnings.
	Customer Assistance, User License (if needed)

Text Block

This block consists of the individual books and topics in the help. Each help book becomes a chapter when a manual is generated.

End Matter

Appendixes that would be included in a printed manual should also be included in the help. Refer to Chapter 6 for some suggestions.

A back cover should be included in the help if a manual will be generated from the help.

Back Cover-Front:	Needed only if a manual is to be generated from the help.
	Company Logo, Product Model Number, Product Name, Manual Name, and Control Number.
Back Cover-Back:	Needed only if a manual is to be generated from the help.
	Company Logo, Company Name, and Address.

Software Examples

Here are some top menus from several fairly well known software programs to show why menu names are not always helpful for organizing a manual or a help system.

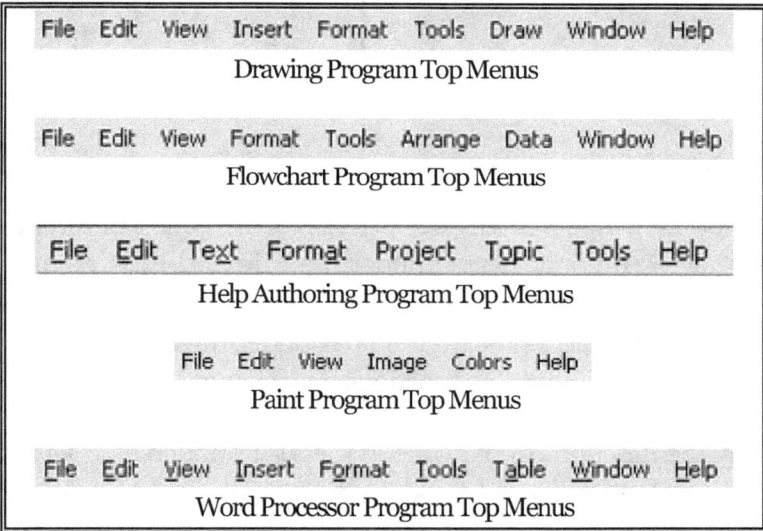

Figure 5–6: Typical Top Menus From Several Programs

Notice that while the top menus are similar, there are differences. The Drawing program has a **Draw** menu, the Flowchart program has a **Data** menu, and the Help Authoring program has a **Project** and a **Topic** menu. These menu differences provide a clue as to the primary purpose of the program

Before looking at some typical help files for these programs, let us look at some of the functions located under a couple common top menus.

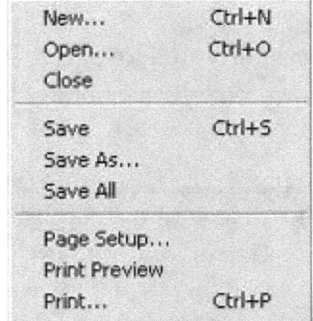
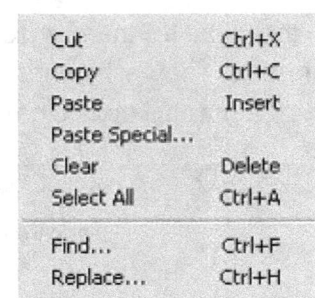

Figure 5–7: Typical **File** Menu | **Figure 5–8:** Typical **Edit** Menu

The top menus shown in the above figures are typical of what functions are commonly located under the **File** and **Edit** menus, regardless of the program. The program's primary purpose determines if other functions are required under these menus and the order or sequence of these functions may change.

Study the functions under the top menus to understand what the program does and what topics or functions should be described and in what order.

Now, let us look at some typical help files for these programs. Help is the primary user document for most software programs and is, therefore, the most important. The chapters of a manual generated from the help should follow the same book and topic organization.

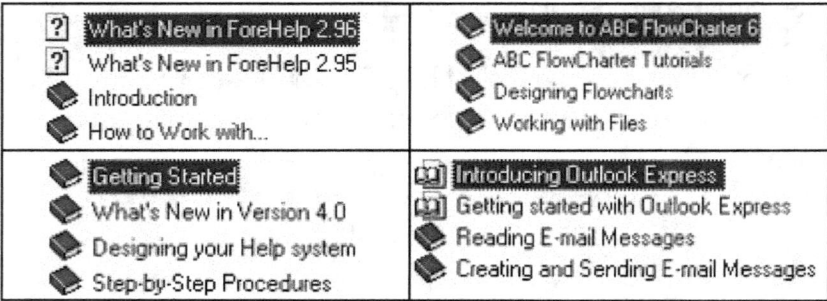

Figure 5–9: Some Typical Program Help Files

NOTE: The graphics in figure 5-9 are from existing help systems and were not created or designed by the author and are presented only as typical examples.

Help files usually start with **What's New** for the latest version, but if this is the first release of the software then a **Welcome to...** or an **Introduction** comes first.

Author's Aside
The best way to learn how to organize help for any software product is to study the help provided by similar products installed on your computer. Study how their help is written and how they present that information to a user.

Help files are generally organized similar to a manual with the exception that software does not require a preface with such things as safety warnings. Also, contact information and licensing information are usually placed last in a help system while these same items are usually placed in the front of a manual.

NOTE: If the user interface follows the "standard" for either the PC or the Mac, a Menu Interface description may not be necessary except for non-standard functions. If the interface is proprietary, describe each menu and each icon.

This Page Intentionally Blank

Chapter 6

End Matter Material

End Matter Overview

End Matter consists of material and information that is not needed to operate and use the product in its intended fashion but may be helpful to the user. End Matter usually consists of a Glossary, one or more Appendixes, and sometimes an Index.

End Matter Material

To help in deciding what should be included in the End Matter, a list of possibilities is provided.

Glossary: This is usually the first item included in the End Matter. Include a Glossary if the manual or help contains 20 to 30 terms that are unfamiliar to the intended audience. A Glossary should also include a list of mathematical or electrical symbols used in the manual. (See Appendix C for an example.)

TIP:　It is a good idea to link a glossary word to its definition, using a pop-up window, whenever it is used in any help topic. This should also be done for any electronic document.

It is a good idea to develop a company master Glossary that includes all terms used in all company manuals. If a manual does not use most of these terms, make a copy of the master Glossary and modify that copy by deleting the unused terms for that manual.

If the Glossary is several pages in length, use headers and footers similar to the ones used in the Text Block. Page numbering should be Glossary specific: G-1, G-2 are examples.

NOTE: There is a trend to consecutively number a manual's pages, especially if the manual will be sent as a PDF.

Appendixes: Include helpful information that an average user rarely needs to access.

Some possibilities for Appendixes are:

> *Product Application Notes*
> *Spare Parts List*
> *Updating Product's software and firmware*
> *How to restore if system crashes*
> *List of error messages and what action to perform*
> *Touch Screen calibration*
> *Product's available options*
> *List of similar company products*

TIP: If the information is needed in a manual, it is also needed in the help.

Appendixes usually contain more technical information about the product and product options. Also, rarely performed procedures and procedures that are only performed by designated personnel may be included.

Each appendix should have headers and footers specific to each appendix, using similar headers and footers as used for the Text Block. Page numbering should be appendix specific: A-1, B-1, etc.

> **Author's Aside**
>
> Changing Heading 1 from Chapter to Appendix and changing page numbering to agree with the End Matter section may be possible with some word processors, using a master document. I prefer separate files for each manual division or occasionally each chapter. It's easier for me.
>
> Also, some writers prefer to number all pages the same way: 1 through 200 or whatever the last page is. I prefer to number each chapter and manual section so as to identify that section: 1-2, A-3, G-1, etc. If you need the total page count, the PDF provides that information.

Index: Consider including an Index if the manual is 150 to 200 pages in length. Word Processors have made the job of creating indexes easier, but by doing so they have made it easier to create indexes that look good but have little value.

If you include an index, do not let the word processor build it without heavy supervision. Delete single words unless they have meaning to the average user. Short two and three word strings are more helpful. If you decide an Index is a must, please study how to properly build effective Indexes rather than just make a list of words.

Better yet, have someone that is proficient in creating Indexes do it.

Author's Aside
Over the years, Indexes have not been much use to me, if the Table Of Contents is well done. I have created several but rarely used them. Besides, if a PDF is created, Adobe Reader can search for a term without an Index and save the time needed to generate what usually ends up being only a fair Index.

The last item a manual needs is a back cover.

Inside Back Cover: Company Logo, Manual Name and Control Number.

Outside Back Cover: Company Logo, Company Name and Address.

Chapter 7

Writing Suggestions

Writing Overview

Technical writing, whether it is for manuals or application help, abides by certain general rules. These rules guide the writer to ensure the information is not lost in a haze of complex sentences and difficult words.

Writing user documentation is different from most other types of writing. You are writing to describe difficult things in such a way as to make them appear simple or at least understandable to inexperienced users. You are not trying to convince the user but rather to inform the user.

Writing Style Suggestions

User documentation is usually formal or semi-formal writing. Over the years, I have found that certain styling and writing rules apply. The following are some of the more important suggestions, regardless of the type of documentation being written.

One possible exception to these suggestions is writing application help, which is usually briefer but with details available if a user wants them. This usually means one topic for the brief instructions or procedures and another topic that details why the instructions or procedures should be followed or performed.

General Style Suggestions

1. Use third person voice.
2. Use active voice rather than passive.
3. Do not sound condescending to the reader.
4. Leave selling to marketing.
5. Use simple, easy to understand words.
6. Avoid long, complex sentences.
7. Write concisely.
8. Write in the present tense.
9. Use past tense sparingly and correctly.
10. Avoid terms like "a lot", "very", "much", "easy".
11. Define terms that may be unfamiliar to the reader.
12. Procedures should be written as if they are being done now (present tense).
13. Describe the work as if you are doing rather than watching.

To expand on some style suggestions:

1. Do not write in a conversational style unless you are writing to a specific person.
2. Active and declarative sentences while good suggest there is only one way to do something or that it has to be done, not could, should, or would be done.
3. Simple words and sentences are best. English may not be the user's first language. A user document may be translated to other languages. Simple words are easier to translate.
4. Use descriptions an average person can understand, unless the document is specifically addressed to engineers and technical personnel.

Manual Style Suggestions

- Divide manual into chapters.
- Page numbers should indicate the chapter: 1-1, 2-3, etc. (Some manuals use consecutively number pages.)
- Use chapter number as part of Figure and Table numbers: Figure 1-1, Table 2-1, etc.
- Do not label tables used only for formatting.
- End chapters on an even page
- Start chapters on an odd page.
- Use mirrored margins for odd and even pages even if manual will be sent only as a PDF (Portable Document Format).
- Place page numbers on the outside of a page. (Header or Footer is usually the best location.)
- Place chapter name on the inside of a page. (Header is better.)
- Use a version number for the manual and for each chapter.
- Avoid using more than four levels of Headings and each Heading should have at least two sub-levels.
- Use typeset characters rather than typewriter characters:
  - Use Em dash — instead of two hyphens - -.
  - Use ° instead of degree.
  - Use "Smart Quotes" instead of "Straight Quotes".
  - Use ½ instead of 1/2.
  - Use 1st instead of 1st.
  - Use µ instead of u for micro.
  - One space after a period — not two.

Heading Example

Page Setup

The formatting of a manual's pages is important. Formatting allows the reader to quickly pick out the Headings and other important information. Heading levels should be visually distinctive so the reader can identify them. This is not possible if all Headings look the same, are of the same point size, or blend in with the text block.

Heading 1 is the beginning of a chapter and includes the chapter number and title, so it should not blend in. Headings 2 thorough 4 should be distinct from each other and from the body text.

TIP: Heading color is not a good visual indicator because some people have trouble seeing certain colors. Using a hanging indent for Heading 2 and 3 is another possibility.

Gutter adds to the Inside margin measurement.

Inside = 0.9 + 0.25 = 1.15

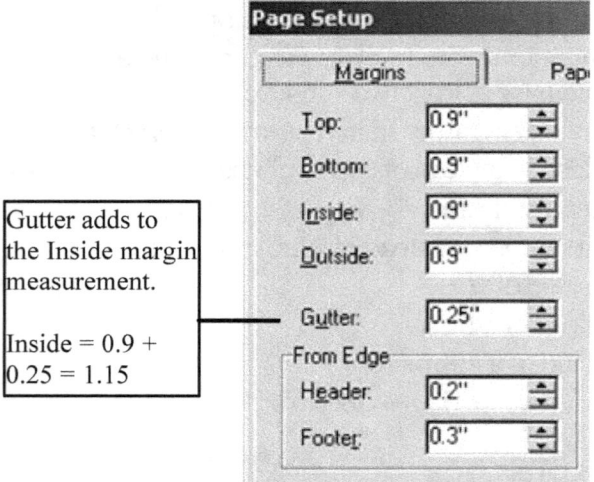

Figure 7–1: Suggested Page margins

Figure 7-1 shows workable margin settings for a word processor document that also includes a Gutter for binding or punching. These Heading suggestions work well for an 8.5 by 11 page but may need modified for smaller page sizes or thicker manuals.

As used in this book, Heading 2 is white text on a black background. This separates Headings 2 and 3 so they are distinctive.

TIP: White space is good if not overdone. A dark, dense page full of type is not inviting to the casual reader.

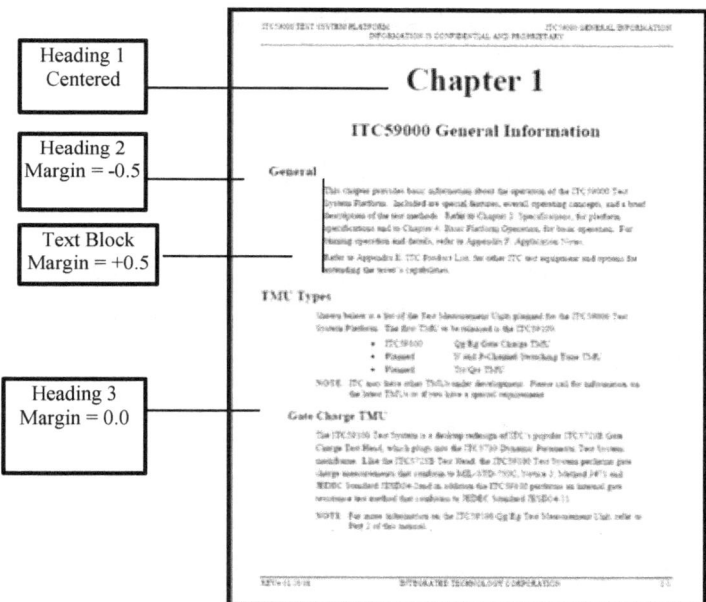

Figure 7–2: Simple Heading Location Diagram

NOTE: See how white space makes a page less intimidating.

Headings	Suggestions
Heading 1:	Chapter Number, Centered, 18 to 42 points.
Heading 2:	First level, Left -0.5, 16 to 20 points
Heading 3:	Second level, Left 0.0, 14 to 16 points
Heading 4:	Third level, Left +0.5, 12 to 14 points
Text Block	Left +0.5, 12 points

Working With Paragraphs

There are four primary paragraph format options: Align Left, Align Center, Align Right, and Justify. The most frequently used option is Align Left and Justified but the others are occasionally used to highlight a sentence or paragraph. Align Right is infrequently used except for tables and special circumstances.

Figure 7–3: Paragraph Formatting Toolbar Icons

Clicking the toolbar icons in Figure 7-3 automatically formats a selected paragraph as indicted by that icon. The following are examples of the paragraph formats and how they might be used.

Here are examples of various paragraph formats:

Align Left:
Align left is normally used for most correspondence. Few books, magazines, and newspapers use this.

Align Center:
Align center is used to call attention to some text. It is often used in place of bold or italics.

Align Right:
Align right is almost never used in print. It may be used in some tables to make some text easier to relate to its label.

Justified:
Use Justify for short line lengths. If the line is more than 4 or 5 inches long, justified text is harder to read than ragged right edge, which is Left Align. Justified is commonly used in publishing.

Hanging Indent, Align Left:
This format is the same as align left but with the first line hanging left of the text block margin. It is used primarily in technical manuals for Heading 2. Lower level headings may be hanging but not as far left.

All formatting covered so far can be included in a style so all that is needed is to select the appropriate style to change a paragraph's format, its font, and its font size in one quick step.

Playing With Fonts

To get familiar with fonts, click the down arrow in the font name box and select other fonts. The best way to compare different fonts is to type a sentence with all the letters of the alphabet and the numbers 1 through 9. Copy and paste this test string several times. Highlight one string at a time and select a different font. The following are examples of some common book and manual fonts.

Arial Font
ABCDEFGHIJKLMNOPQRSTUVWXYZ
abcdefghijklmnopqrstuvwxyz 1234567890

There are several variations of Arial from bold to narrow and each provides a different look. Arial works best as Header or Footer text, headings text, and as book title text. I don't recommend it for block text except for occasional use.

Bookman Font
ABCDEFGHIJKLMNOPQRSTUVWXYZ
abcdefghijklmnopqrstuvwxyz 1234567890

Bookman is a nice serif font. It works best as Header or Footer text, Headings text, and as title text. Letter spacing is good but lower case is small. It could be used at 13 points for better readability.

Century Schoolbook Font
ABCDEFGHIJKLMNOPQRSTUVWXYZ
abcdefghijklmnopqrstuvwxyz 1234567890

Century Schoolbook is a serif font. It works best as Header or Footer text, headings text, and as title text. Lower case letters are close and small.

Garamond Font
ABCDEFGHIJKLMNOPQRSTUVWXYZ
abcdefghijklmnopqrstuvwxyz 1234567890

Garamond works best as Header or Footer text, headings text, and as title text. Lower case letters are close and small.

Georgia Font
ABCDEFGHIJKLMNOPQRSTUVWXYZ
abcdefghijklmnopqrstuvwxyz 1234567890

Georgia is a serif font and works best as a manual's block text. Lower case letters are close. Letters and numbers can look odd together because some numbers extend below the base line. This is the font used for the text of this book.

Helvetica Font
ABCDEFGHIJKLMNOPQRSTUVWXYZ
abcdefghijklmnopqrstuvwxyz 1234567890

Helvetica is a sans-serif font. I don't have much use for this font but some people love it.

Palatino Font
ABCDEFGHIJKLMNOPQRSTUVWXYZ
abcdefghijklmnopqrstuvwxyz 1234567890

Palatino is a serif font and works best as Header or Footer text, headings text, and as title text. Lower case letters are close and small.

Times New Roman Font
ABCDEFGHIJKLMNOPQRSTUVWXYZ
abcdefghijklmnopqrstuvwxyz 1234567890

Times is a serif font and works best as block text at 13 points. Lower case letters are close and small.

Trebuchet MS Font
ABCDEFGHIJKLMNOPQRSTUVWXYZ
abcdefghijklmnopqrstuvwxyz 1234567890

Trebuchet is a sans-serif font and works best as Header or Footer text, headings text, and title text.

Verdana Font
ABCDEFGHIJKLMNOPQRSTUVWXYZ
abcdefghijklmnopqrstuvwxyz 1234567890

Verdana is a sans-serif font and works best as Header or Footer text, headings text, and title text. This is the font used in this book for title, Headers and Footers, and for headings text.

NOTE: These examples are set to 12 points, justified, and with identical line spacing so you can compare them. Notice the physical size and readability of each font.

Font choice is a personal preference as to which one appeals to the writer's eye. A font should be selected based on its readability, so test before committing to a font. Also, remember all your readers may not have 20/20 eyesight, so select a font size that's easy to read. This is usually no smaller than 12 points.

NOTE: Do not type all CAPITAL letters, use *Italics* only for attention, use **Bold** sparingly, and avoid <u>underlining</u> compete paragraphs.

This Page Intentionally Blank

Chapter 8

Getting Started

Getting Starting Overview

Getting started on a new manual or any new user document can sometimes cause writer's block, so having an established starting method can help break the brain cells loose. This chapter also provides some suggestions for interviewing the design team.

In technical writing, the subject of a manual is not debatable but the manual's contents are. "Here is the product; where is the manual?" is the statement bosses tend to utter. The product is defined or will be soon, so how to get started on that manual is the first step.

Some of the suggestions in this chapter may be found scattered throughout this book, but as any good manual, interviewing and "How To Get Started" suggestions should be located in one place.

NOTE: Some suggestions in this chapter may not apply to all products and all user documents.

Interviewing Suggestions

Interviewing is an art and a skill that requires time and practice to master. It also requires the knack of extracting information from busy and sometimes reluctant interviewees without being thrown out of their work area—too often.

The more you know about the product and the person to be interviewed, the better chance you have of obtaining information without being too annoying, unless you enjoy that part.

There are two basic types of interviews: face-to-face, and telephone. The face-to-face should be used primarily for in-depth interviews while the telephone should be used for clarifying a specific detail covered in a previous face-to-face interview.

Some Interviewing Tips

1. Make an appointment if the interview will take more than 15 to 20 minutes.
2. Know as much as possible about the product.
3. Make a list of questions that you want answered.
4. Record the interview.
5. Ask, then listen.
6. Ask questions that require detailed answers.
7. Show interest—do not bore the interviewee.
8. Keep the interview focused on the product.
9. Do not pretend to know more than you do.
10. Ask for clarification as needed.

The art in interviewing is to appear interested even when you can only understand every third word. The skill is in not letting the interviewee know you are in over your head, so here are a few tricks:

- Nod and say "I see" or "I understand" or "OK".
- Keep the interview on target even if you are lost.
- Do not waste interviewee's time.

Start the interview with simple questions to get a feel of how the interviewee reacts. Ask leading, detailed questions that require the interviewee to explain and expound. If you do not understand something, say so. A simple, "Can you say that simpler." works wonders. Usually the interviewee is happy to go into more detail.

However, if most of what the interviewee says is going over your head, stop the interview and do some research until you understand the basics of the product. If you appear lost, the interviewee will not be inclined to waste his time telling you what you do not or cannot understand. By all means, try to look intelligent. Show puzzlement: raise an eyebrow or frown and hope your timing is right.

> **Author's Aside**
> Even if you try NOT to be annoying, you will usually be considered as such, so get used to it. Take pride in getting the information rather than in being annoying; that is just a side benefit.

Gathering Information Suggestions

In most cases, your boss will say, "We have a new product and we need a manual. It's not quite done yet, but it will be finished in about a week." The timetable will probably be off by a few weeks or a few months, but lack of a real product is not a major problem to a good technical writer. (This is where creativity comes into play.)

A good writer often waits for inspiration to strike but the boss may not appreciate seeing your feet on the desk and your eyes closed. Writers know this as the planning stage; however, your boss may take a dim view of this stage and strongly suggest you plan with your eyes open, feet on the floor, and a pencil or a keyboard in your hand. Most bosses, unless they are also writers, do not understand the value of this planning and plotting stage. They just assume that if you are not active, you are loafing.

Try to look busy. Equipment needed is a notepad and possibly a tape recorder. At this stage, the recorder is more of a "just in case" thing, but the notepad makes you look as if you are not just wandering around bothering people. Remember to jot down a few notes every now and then so you appear to be attentive.

Now that you have some idea how to extract information from the design team, it is time to confront them and start gathering information. Remember, while interviewing the design team, you are representing the potential user and customer.

Starting Steps
1. Talk to the programmer
2. Talk to the design engineer
3. Talk to marketing
4. Talk to technicians working on the product
5. Product's Information review
6. Decide on manual types

Finally, remember to tread lightly in unfriendly territory. Engineers and programmers do not like having their work critiqued until they feel the project is near completion. They may also be feeling deadline pressures. They do want to help, but they do not like disclosing information until they are sure it is correct, much like writers.

Talk to the Programmer

Start with the programmer. The software may not be done but some screens may be. Ask the programmer if the software has a simulation mode or if it can operate alone. If not, ask if the programmer can give you a copy of the screens that are somewhat done.

NOTE: The programmer is the primary source for a software product other than you learning to use that product.

Good programmers often provide a simulation mode. Simulation mode allows you to review the product's interface operation before the product is available and without bothering the programmer. (Programmers do not appreciate being interrupted.)

TIP: Simulation can also be used to preview the product to customers, and later it can be used to help train users without damaging the product.

Talk to the Design Engineer

Now that you have either a beta of the software or a few screen shots, talk to the product's hardware designer. The designer can give you an overview of what the product does or is supposed to do. Ask about what parameters are needed for input, the size and shape of the product, and is it bench or rack mounted?

Talk to Marketing

Check with Marketing; they should be able to add to the product's description. There may be a customer's design specification that provides more information about the product. Specifications, such as what voltages are needed and what tests and what test parameters are required, can help you understand the product.

Talk to Technicians

Talk to anyone and everyone. Do not overlook the technicians. They usually have an insight into the operation of a product at a level of detail that designers and programmers often overlook.

Product Information Review

Remember, you should be looking at the product from the user's and customer's point of view. If something does not look or feel right, talk to anyone that will listen and anyone who might want to or can fix that problem.

As you look over the screens, verify if all inputs have range checking, especially inputs such as voltages, currents, and test times. If a test voltage input has a range of 1 to 100 volts, what happens if zero (0) or 101 is input? What is the step increment: 1 volt or 0.1 volt? Is the voltage input in volts: 1.0 or millivolts: 1000.0?

Decide On Manual Types

As you begin to understand the product, you must decide what type manual the product requires. This decision may not be yours to make, marketing may make the decision for you or the customer's purchase order may determine at least the name of one of the manuals.

The final decision of the manual type may be based on the product. If it is primarily software, that manual will be different from a hardware-firmware combination product.

The output format of the manual should also be considered. If the manual is primarily done as help and then converted to a PDF or word processing document, writing the help topics may come first and then the organization of those topics into chapters may come later. (See Chapter 13.)

Another source of information may be the Internet. If the product is new to you, there may be similar products on the web that can give you some ideas of where to start.

Creating Manual Outlines

Before starting a new manual, it is a good idea to create an outline that shows the manual's basic chapters and their intended contents. An outline is a good way to get started and it helps to engage the mind. There are three basic types of manual outlines: Software only, Hardware only, and a Hybrid, which is a combination of both.

All outlines are similar, only the positioning of the information within the outline differs. That is, most manuals have similar Front Matter; although, the type and amount of information changes based on what the product does and the manual type.

Example Manual Outline
This following example hardware/software outline is incomplete but it shows how to start an outline for most hardware-software manuals. An outline is usually requested so the boss knows what you intend to do. An outline also shows if you have a good grasp of the product or are blowing smoke.

Front Cover
Warranties
Software Disclaimer
Copyrights
Trademarks

Preface (or About This Manual)
Customer Assistance
Proprietary Notice
Manual Overview
How To Read This Manual
Manual Revisions
User License
Safety Requirements

Table Of Contents
To Be Determined (Determined by the text block)

Quick Start (optional)
A list of things needed to start product in production operation:
Use small screen and box graphics
Refer to manual chapters for more detail
Include:
Turning product on
Loading of software
Creating part specifications
Adding part numbers

General Information
Product Overview
Product Operating Modes
Production Mode
Engineering Mode

Chapter 9

Writing "How To" Procedures

Procedures Overview

All technical manuals should have procedures that describe how to perform tasks the user may not understand or has little experience at performing. Provide a list of procedures in the Table Of Contents and make all procedures available in online help so the user can access a procedure while operating the product.

Procedures Tips

When writing procedures, it is important to be consistent. Here are a few suggestions:

Pre-Instruction Suggestions:

- List the tools and equipment required for a procedure.
- List special training or knowledge needed for a procedure.
- Label a procedure so there is no doubt as to what it does.
- Place warnings before a procedure.
  Example: **DANGER:** High Voltage.
- If an instruction requires unusual conditions, place a warning before that procedure and also at that instructional step.

Instruction Suggestions

- Start each instruction with a verb. Examples: Unplug power, Close valve, Turn power switch off, Remove screws

- If covers are removed, indicate the number of screws to remove. Example: Remove the 6 screws holding the top cover.

- Number instructions to indicate their sequence.

- Place only one action in an instructional step.

- If performing an instruction results in a reaction, indicate what reaction to expect. Example: Remove power. Wait for capacitors to discharge.

- If dangerous voltages are possible, add instruction to discharge voltages.

    NOTE: Necessary if storage capacitors are in the product.

- Show and tell what is expected as each instruction is done.

- Write short, concise instructions but be sure that an inexperienced user understands the instructions.

- Assume the user is unfamiliar with the product and with the procedure.

- Keep procedure steps to 10 or 15 simple steps.

- Create one or more sub-tasks if a procedure exceeds 15 or more steps.

- If covers were removed, include replacing those covers.

- When procedure is complete, indicate what user should do next. Example: Reconnect power, Turn power switch On, etc.

- Specify what the user does next. Example: Resume Testing.

Procedures Examples

Procedures can be presented in several ways, so to help with this, here are five procedures formatted in different ways to help you see the different possibilities.

Procedure To Remove a Board

How To Remove The B1 Board (Example 1)

Equipment Required

- Cross-point, plastic handled screwdriver
- Static wrap to protect board
- Ground strap

> **!!! WARNING !!!**
> Dangerous voltage on C-1 Capacitor.
> Discharged before removing B1.
> See Step 6.

Instructions To Follow

1. Turn Power Switch Off.
2. Unplug power cord from product.
3. Attach ground strap to wrist and to chassis ground.
4. Remove the 4 screws securing the top cover.
5. Remove cover and store.
6. Temporarily short capacitor C1 across terminals with plastic handled screwdriver. (C1 may arc.)
7. Lift tabs on both ends of board B1 to unseat the board.
8. Remove board B1 and place on static guard material to prevent damage.
9. Insert new board B1 and press down to seat.
10. Replace cover removed in step 4. (4 screws)
11. Plug in power cord.
12. Turn Power Switch On.
13. Wait for product to start up; resume testing.

The above procedure is a simple list with no pictures or comments. For simple, easy to understand procedures, this is usually sufficient, but as example 2 shows, there may be a better way.

How To Remove The B1 Board (Example 2)

Equipment Required

- Cross-point, plastic handled screwdriver
- Static wrap to protect board
- Ground strap

> **!!! WARNING !!!**
> Dangerous voltage on C-1 Capacitor.
> Discharged before removing B1.
> See Step 6.

Removing The B1 Board		
Step	**Instruction**	**Comment**
1	Turn Power Switch OFF.	
2	Unplug power cord from product.	
3	Attach ground strap to arm and chassis ground.	Product is static sensitive.
4	Remove the 4 screws securing the top cover.	Save screws in safe place.
5	Remove Cover and store.	Store in safe place.
6	Temporarily short C1 across terminals with plastic handled screwdriver.	C1 may spark and arc when it discharges.
7	Lift tabs on both ends of board B1 to unseat the board.	Board should be loose in its slot.
8	Remove board B1 and place on static guard material to prevent damage.	
9	Insert new board B1 and press down to seat.	Tabs on board end should be flat.
10	Replace cover remove in step 5.	Use screws removed in Step 4
11	Plug in power cord.	Power switch must be OFF.
12	Turn Power Switch ON.	Wait for product to start up; resume testing.

The above procedure is the same as example 1 but uses a 3-column layout. Pictures could be added for more help.

Procedure To Log In

How To Log In (Example 1)

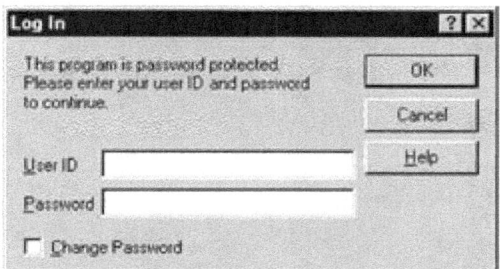

Figure 9–1: Typical User Log In Dialog Box

Instructions To Follow

1. Place cursor in **User ID** field.
2. Type user's ID. (Assigned by Administrator.)
3. Place cursor in **Password** field.
4. Type password assigned to the user.
   (User can change password after logging in.)
5. Click **OK** to log in the user.
6. Product's Start-Up screen is displayed.

Because this is only a log-in procedure, the other buttons and the Change Password checkbox are not described. Those items should be described in a **Getting Started** or **Basic Operation** chapter where the log in dialog box is described in detail.

How To Log In (Example 2)

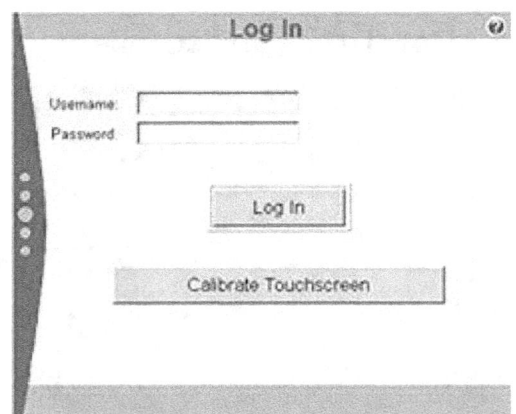

Figure 9–2: Another Example Of A User Log In Dialog Box

Instructions To Follow

1. Place cursor in **Username** field.
2. Type user's name. (Assigned by Administrator.)
3. Place cursor in **Password** field.
4. Type password assigned to the user.
   (User can change password after logging in.)
5. Click or touch **Log In** to log in the user.
6. Product's Start-Up screen is displayed.

How To Log In (Example 3)

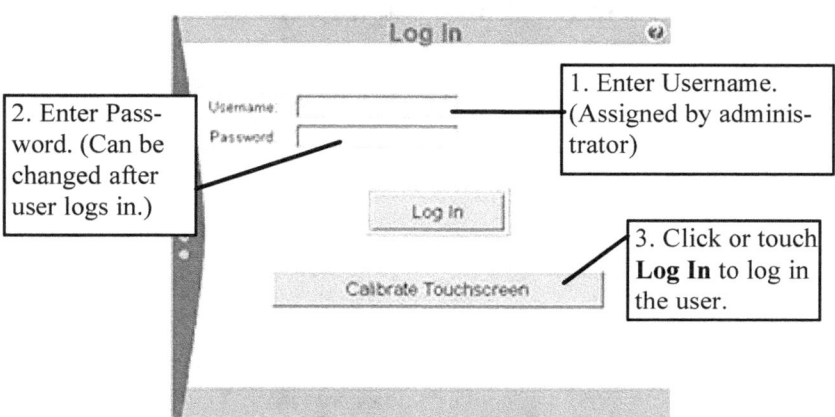

Figure 9–3: A Last User Log In Example

The above example shows a simple procedure, using only the space required for the graphic. This type procedure can be effective but only for short and simple procedures.

The best use of this technique is to define screen functions without having to look elsewhere for that function's description. Again, this technique works best on simple screens and dialog boxes with few functions.

This Page Intentionally Blank

Chapter 10

Working With Graphics

Because manuals need graphics, you should know how to capture and manipulate those images to achieve the best results. If the manual is intended for printing, images should have a PPI of at least 200. But if the manual is primarily intended for monitor viewing, the PPI does not need to be more than about 100.

This chapter describes how to change the PPI of an image, using Photoshop, and how to manipulate and save that image so it does not degrade when the manual is converted to a PDF.

Photoshop Basics

Let's start with the very basics of working with Photoshop. While this book will not teach you all you need to know about Photoshop, it does tell you enough to help you create a quality manual, regardless if its intended or primary use.

In most cases, you will capture an image from your monitor of a dialog box or screen that needs to be described in the manual. This image can be directly pasted into Photoshop and then the resolution increased and the image sized for best viewing, based on the manual's page size.

Opening A New Canvas

Photoshop refers to a new page as a workplace or as a canvas. So the first thing to do is open a new canvas.

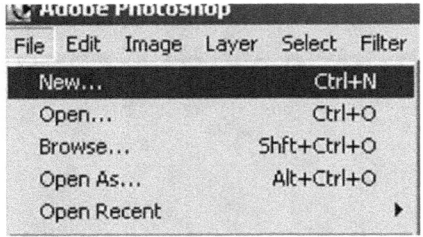

Figure 10–1: Photoshop's File Menu

When **New** is clicked on the **File** menu, the **New** box is displayed.

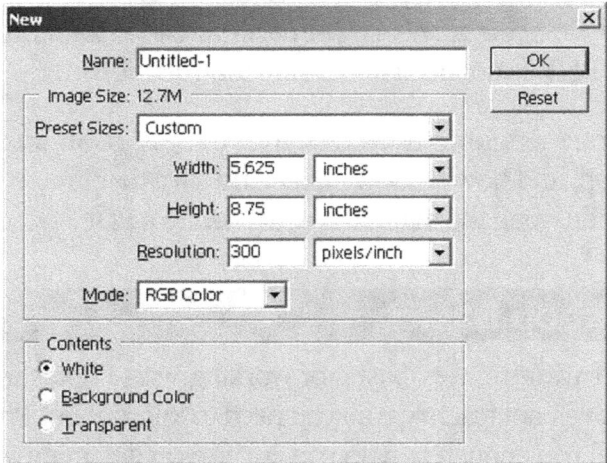

Figure 10–2: The New Dialog Box

If you have captured an image, the **New** box automatically shows the size and the resolution of that captured image. Click **OK** to close the **New** box and display a canvas of the size and resolution shown in the box. Paste the image on to the new canvas.

You can name the graphic before clicking **OK** or wait until you have manipulated the image and want to save it.

Photoshop's Toolbar

Photoshop's main toolbar is used the most when working on an image. Only a few toolbar items are necessary to change an image for use in a manual.

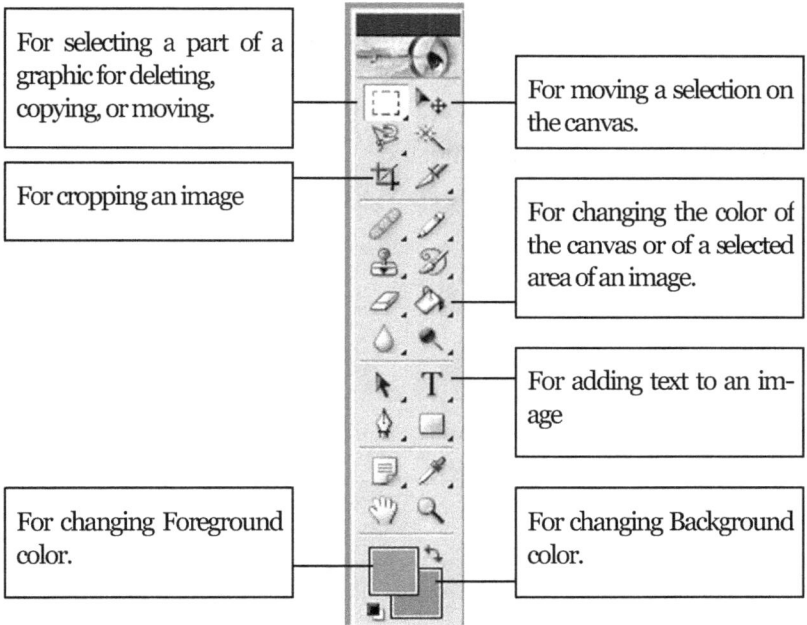

Figure 10–3: Photoshop's main Toolbar

Working With Images

The captured image must be adjusted in Photoshop for the size you intend to use in your manual. While JPEG files do scale well, it is better to do most of the major size adjustments in Photoshop rather than in Word or any word processor.

When working with images captured from a monitor for use in a manual, you face two conditions that require different approaches for best print quality.

1. The image is close to the desired physical size but the reso-
    lution is too low for quality printing.

2. The image is too large for the manual and the resolution is
    too low for quality printing.

Condition 1 Example

After a captured image is pasted into Photoshop, click **Image Size**
on the **Image** menu to display the **Image Size** dialog box. If the
image is near the correct size but the PPI are less than desired, only
the Resolution needs to be adjusted.

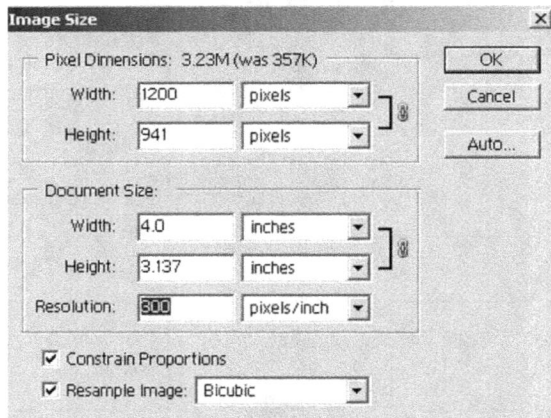

Figure 10–4: Changing The Resolution Of An Image

Because the above Image Box was captured from a monitor, its
resolution is 72 PPI, which is too low for good printing. To change
the PPI to a value suitable for printing, enter the desired value in
the **Resolution** field. (I entered 300 PPI.)

Do not uncheck **Resample Image** for this example.

If the manual is primarily for viewing on a monitor but it might also
be printed, a Resolution of 150 to 200 PPI should be sufficient. If
the manual is intended primarily for printing, the Resolution
should be 200 to 300 PPI.

By increasing the resolution to 300PPI, the image is readable even if reduced by 50 percent in your word processor.

Condition 2 Example

The captured image is larger than what is needed for the manual and its resolution is too low for printing, the size needs to be reduced and the resolution increased.

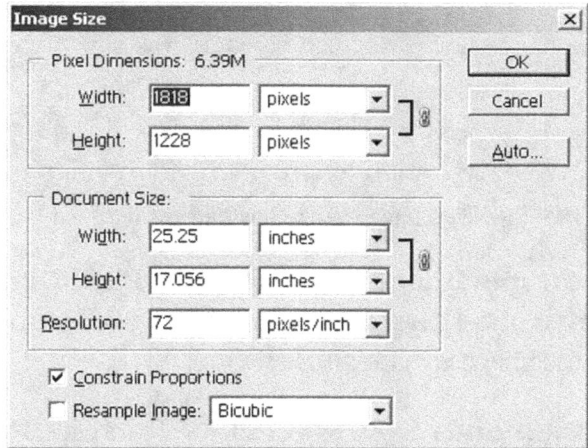

Figure 10–5: Image Size Before Adjustments

To do this without degrading the image, uncheck the **Resample Image** checkbox (See Figure 10-5). This allows the image to be made smaller without adding pixels to or subtracting pixels from the image.

Leave **Constrain Proportions** checked as this allows the image's size to be changed without distorting it. As Figure 10-6 shows, the image's **Width** is set to 4.5 inches. Notice the new **Resolution** value and the new **Height**.

If the **Resolution** is not high enough after the image is reduced in size, you can click **Resample Image** and change the resolution to a desired value.

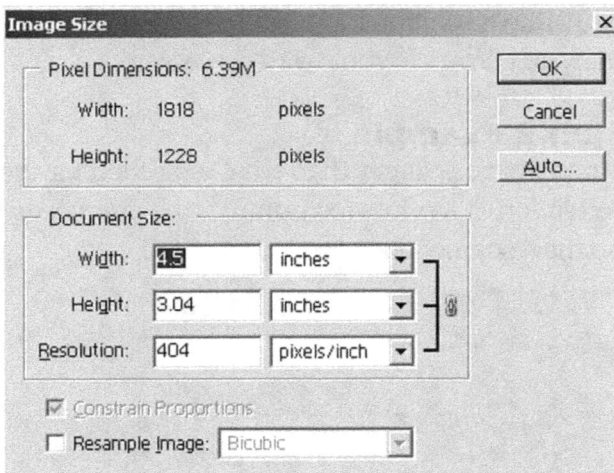

Figure 10–6: Resized Image With New Dimensions And Resolution

When **Resample Image** is unchecked, the **Pixel Dimensions** area is grayed out. This is because the actual pixel count does not change as the image is made smaller.

If the **Resolution** is above 300 PPI, you can click **Resample Image** and then enter 300 or the desired PPI before clicking **OK**.

Click **OK** to adjust the image to the values shown in the **Image Size** box of Figure 10-6.

Saving An Image

After you have changed your image to the desired size and resolution, that image must be saved with minimal changes to its resolution, especially if the manual will be printed. I suggest you work with JPEG files because they appear to be best for normal manual printing.

To save your image, click Photoshop's **File→Save As** to display the **Save As** box. Select the JPEG format to save to. Change the name of the image or save it to the same file name with JPG extension by clicking **Save**. When **Save** is clicked in the **Save As** box, the **JPEG Options** box appears.

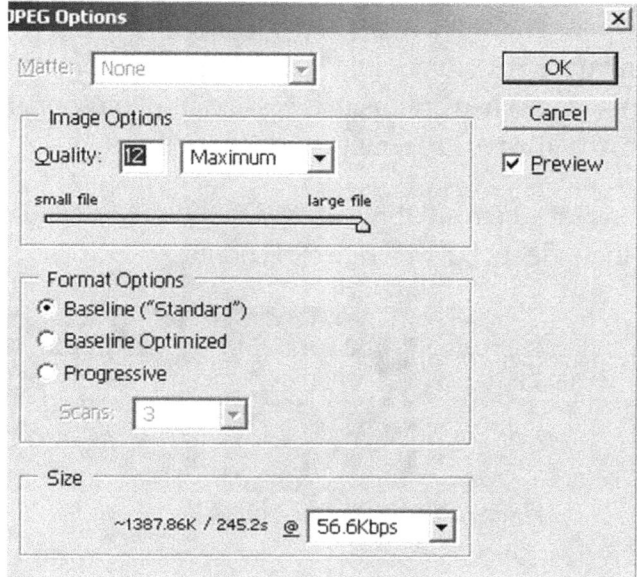

Figure 10–7: Setting The JPEG Save Options

Notice the **Quality** is set to **Maximum** or for large file size. It is better to save your image with minimum compression. This is because when a PDF is created, you may need to compress the file so its file size is acceptable for downloading. If the manual is shipped on a CD, the file size may need to be adjusted to fit on the CD.

NOTE: If saving the image for the web or for monitor viewing, use **small file** size or **minimum** file size.

You get better results letting Acrobat compress the file once rather than compressing the image in Photoshop and then compressing it further when the PDF is created.

Click **OK** to save the image.

PPI Problems

The following demonstrates a problem that can occur when working with images.

When an image is imported into Word, its physical dimensions determine its size. When an image is created in Photoshop, its size is relative to the image's PPI count.

To demonstrate, here are three images that have the same physical size of 1-inch square but different PPI counts.

Figure 10–8: Images Imported Into Word

Figure 10–9: Images Combined In Photoshop

This shows that if a 100 PPI image is pasted into a 300 PPI image in Photoshop, the actual size is based on the image's PPI not the physical size as stated in the Image Size box.

Chapter 11

Online Documentation

Online Documents Overview

Online documentation covers a wide range of possibilities: eBooks, PDFs, WinHelp, MS HTML Help, JavaHelp, and several other types including videos and interactive tutorials. The only thing electronic documents have in common is they are viewed on a computer. Some require a special reader or an Internet browser.

The trend is toward more online documentation with online help becoming the main delivery method. Most help authoring programs allow the help to be compiled as either HTML help or as WinHelp for legacy products. Also, most help authoring programs allow the help to be outputted to a word processor format, to a PDF, and some even to an eBook or a Pocket PC format.

If there is a need or desire to create an eBook, most eBook programs import HTML files so there is no major work needed to create just about any type electronic documentation wanted.

Online Documents Problems

Online documentation has one major problem: A computer is required to read it. This usually means the product must be at least partially operational to view the online documentation. A paper manual requires no power

If the user knows that an action has to be performed after the product's power is removed, a copy of that procedure could be printed prior to removing power. But, what happens if the product crashes or suddenly fails? Most products do not warn that a failure is about to happen. It would be nice if all procedures were available for off-line use, or at least those procedures that require the product to be powered down.

A second problem is the average person dislikes reading long documents on a monitor or display. The answer is to create short documents rather than one long document or create the document with short paragraphs, short chapters, and easy to remember procedures.

NOTE: Because help uses individual topics, it is an ideal method of displaying information to users without giving them more than they need. (If the product is operational.)

Online Documents Suggestions

To counter the problem of performing procedures when the product is turned off, those procedures could be taped or glued to the product's case. Of course, this only works if the product is larger than the document that has the procedures.

Procedures could be placed inside the product so, if a cover is removed, the required procedure would be available. But if removing the cover requires a procedure, this idea does not work.

A better solution is to provide a display that is easily removed and that operates when the product is powered down. A touch screen display with internal memory and a CPU (Central Processor Unit) that is battery powered and runs a small, simple application, is one answer. This self-operating and self-contained display provides access to procedures when the product itself is not operational. This would allow the user to move the display to where the work or adjustment is needed.

This same display could also show the installation procedures before the product is installed. Of course, once the product is installed and power is applied, the product powers the display and its internal battery is recharged. When the product is operating, the display shows the product's user interface.

Another possibility is to allow some or all procedures to be copied to smartphones and PDAs. Many users have smart cellphones that use the Microsoft Windows Mobile operating system. Some help authoring programs can save help files to a Pocket PC or an eBook format that run on most Windows powered cellphones.

Online File Types

There are several very popular file formats for online documents. The leader appears to be the PDF (Portable Document Format) by Adobe; however, there are less expensive programs that create PDFs. A less expensive program may not have some features that Adobe Acrobat offers.

Another popular type of online document format is the eBook. This format is becoming very popular for authors wishing to publish without a large up-front outlay. Also, eBooks do not require printing and can be easily downloaded on the Internet.

PDF Suggestions

A PDF is the best way to ensure that what is seen on your computer (format, fonts, and all) looks exactly the same way on all other computers if the PDF is correctly created. A PDF also ensures that once created and the file delivered to a printer or a customer, the file will print exactly as you formatted it.

This means a technical writer should follow some basic rules and steps when creating a PDF of a manual.

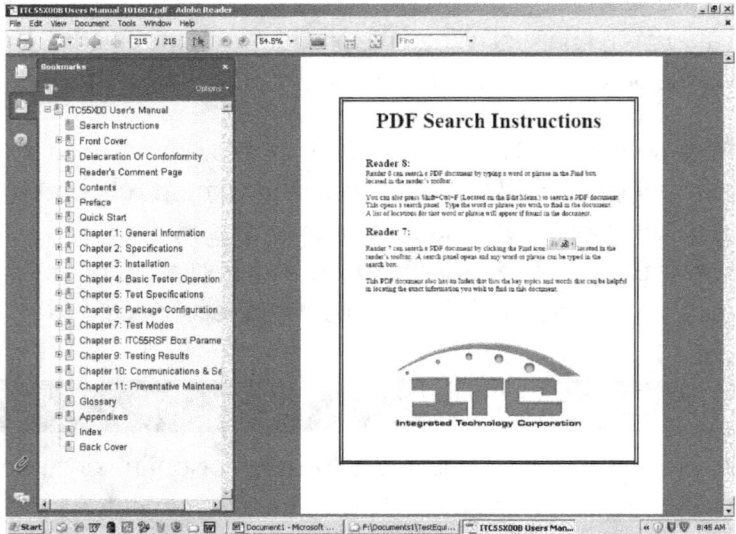

Figure 11–1: Acrobat Reader With Navigation Panel

In order to navigate to the desired chapter or paragraph within a chapter, the manual's Headings should be used to create book-

---

marks in the PDF. These bookmarks allow the user to click on them and be taken to that location in the manual (See Figure 11-1).

There are two ways to create bookmarks for a PDF: manually or automatically. Manually means the writer inserts the bookmarks for each of the chapter headings and verifies that each is correctly linked.

Automatically bookmarks can be created from Word. By clicking on the Create PDF icon or the PDF file menu function, each chapter's headings are added to the PDF and bookmarks and automatically linked to the appropriate location. (Heading 1 becomes a book and the sub-headings are indented and placed in that book.)

There is one possible problem with the automatic PDF method: all fonts used in the manual may not be embedded in the PDF. Another possible problem is that when printing to custom page sizes, verify that the PDF output page size matches the document page size. If you are printing a 5.5 by 8.5 page, it may come out as an 8.5 by 11 instead.

What this means is test before sending a PDF to anyone to make sure that what he receives is exactly what you intended.

eBook Suggestions

Another possible choice for online documentation is the eBook. While not as flexible as PDFs, eBooks may be a good idea for small manuals with few images. eBooks allow buyers to download and read within seconds if not minutes—instant gratification but so do PDFs.

Some eBook authoring programs offer security that is specific to book publishing. This includes security that allows the eBook to be viewed only on the computer it is initially installed so the file cannot be passed around. The reader cannot modify the file and the displayed format can look like a book if you want.

Some of these security features are also available with PDF creation programs but the document must first be created in a separate program and then converted to PDF.

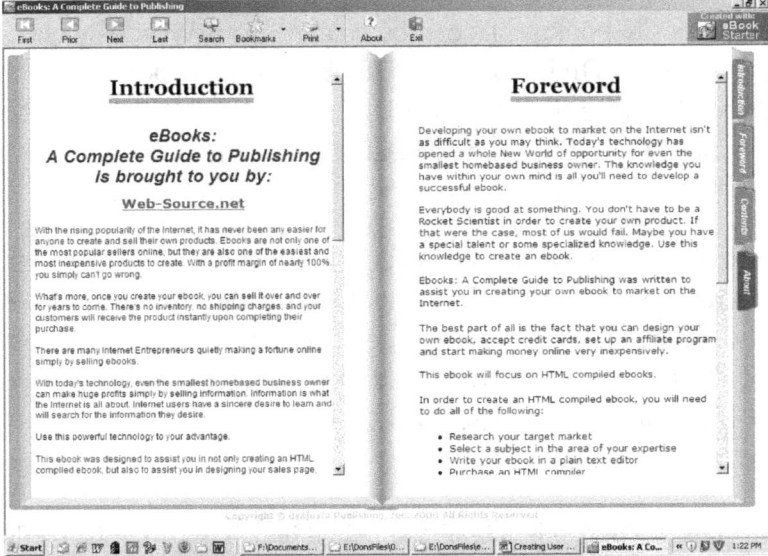

Figure 11–2: eBook Example (eBookStarter.com)

Before deciding on the eBook approach, make sure it allows your customers to print all of a page. This problem can be fixed, in most cases, by not allowing the text to exceed the book's physical page size. (No scroll bars) As the above example shows, scroll bars are used to display additional text on the page but that text does not print.

Most eBooks are similar to websites and allow for hyperlinks to topics in the eBook. This also means that if you are familiar with HTML, creating an eBook is not that difficult. Plus, some eBooks allow the document to be created in the program, similar to some help authoring tools. Some eBooks import HTML files and graphics and operate much like a word processor.

One note of caution, there are many eBook-authoring programs available with prices that vary from less than one hundred to several hundred dollars. If you are interested in trying this route, start with *http://www.ebookresourcecenter.com:80/*. This page compares some eBook programs and offers helpful information.

This Page Intentionally Blank

Chapter 12

E-learning

What Is E-Learning?

E-learning is often referred as distance learning. It is usually classed as any learning that uses technology to assist or present a subject, especially on a computer. This means any subject that can be committed to a video or a screen recorder or that can run on the web or from a CD or DVD can be called e-learning. However, e-learning usually applies to an individual learning experience rather than to an organized classroom experience.

Learning, any learning, can be classed as e-learning because most subjects can be developed and adapted for running on a computer.

Why We Learn

We must realize that regardless of the quality of the course or of the instructor, a student will not learn unless he has:

- The emotion and the motivation to learn,
- The responsibility and the need for learning, and
- The persistence and the patience to learn.

If these factors cannot be awakened in a student, he will not learn. So how to make a student aware of his part in the learning process? First, the student must want to learn and must be given a reason to learn. The best motivation for learning comes from the student's character, not from external rewards or pressures.

This could be as simple as telling a student that his advancement depends on knowing the subject, appealing to his self pride, or a pay raise may depend on how well he understands the subject. Without motivation most students learn the minimum possible to scrape by. Simply telling a student he ought to learn or it is his duty to learn will not cause him to learn. He must want to learn.

> **Author's Aside**
> The design of a course can encourage a student to learn but it can never make a student learn. Using course phases that must be passed before the next phase is started, is one method.
>
> **Course design is an art—not magic.**

How We Learn

In order to help a student learn, we must understand how he learns and then decide if the methods we are using are addressing these proven learning methods.

Experts say we learn through our senses, but how else could we learn. While taste and smell are helpful in leaning some things, most of our learning occurs though hearing, vision, and doing.

Hearing

This type student, which is often referred to as an auditory learner, learns best by hearing the information. He usually finds written information difficult to grasp. This type student benefits from reading the information aloud and by taping a lesson so he can hear it later. He often gains information by listening to how the instructor speaks. Tone conveys which information is important.

Vision

This type student uses his eyes to gather information. His learning is assisted by such visual aids as charts, diagrams, tables, illustrations, and videos. This type student learns best in a quiet environment with few interruptions. He often takes notes or likes to markup the material as he reads.

One problem with this learning method is the student may prefer reading to watching videos. Also, a younger generation may not have the patience to read because they are used to watching TV and playing video games. Older students may be more open to reading if they grew up entertaining themselves with books instead of video games.

Doing

This type student is usually referred to as a motion and touch learner. This student learns by doing, which amounts to muscle memory: which is a hands-on learning method. This student does not like watching a video that shows him how to perform a task, he prefers to do the required actions even if only in simulation mode. This student learns best by performing, not by watching or being told or even by reading.

Think of the things you have learned. Things like driving and golf. Learning the rules was necessary but the doing was what it was all about. Usually, doing is more fun and educational than watching.

Documentation Suggestions

Why are these learning methods important; because these methods must be considered when developing user documentation. Here user documentation refers to manuals, courseware, marketing material, or any document that ends up in a user's or customer's hands to describe or promote a product.

Because people learn and gather information differently, user documentation should be delivered in different formats so they can select the format most appropriate to them. The different formats allow them to read, hear, or see the information they want in the manner that is most meaningful to them.

While it may not be possible or feasible to deliver all user documents in different formats, you should consider doing so for the most important parts of a manual, as an example. What is most important should be determined by the student, user, or customer, not the instructor or writer.

Each of these formats should say similar things but not in the same way. A video should show more than the printed brochure and the video could also contain audio so it can be listened to even if it is not watched.

Paper manuals are becoming obsolete but what is the most appropriate replacement? True, HTML help is a major contender but it alone cannot satisfy the requirements for each type learner. Yes, you can add video and audio recordings to fill the gaps but help is usually written for and directed at a specific application and not intended as a standalone application or document.

With some imagination and creative thinking HTML help can be used to create product demos, tutorials, and even basic courseware.

You can also create demos that use only HTML, CSS, and a little JavaScript. But, these methods lack the flexibility of programs such as Adobe's Captivate and Techsmith's Camtasia that are specifically designed for creating e-learning applications.

The programs and the learning methods you use determine what you are able to create. As one example, can you create an interactive application or demo that encourages the student to follow audio and visual directions but he must do the actual steps himself?

I believe the most effective tutorial or product demo must not just show the student what should be done but should encourage him to click a button, type actual inputs, and use the demo or tutorial as if simulating the actual product or at least some if its functions.

Some Examples

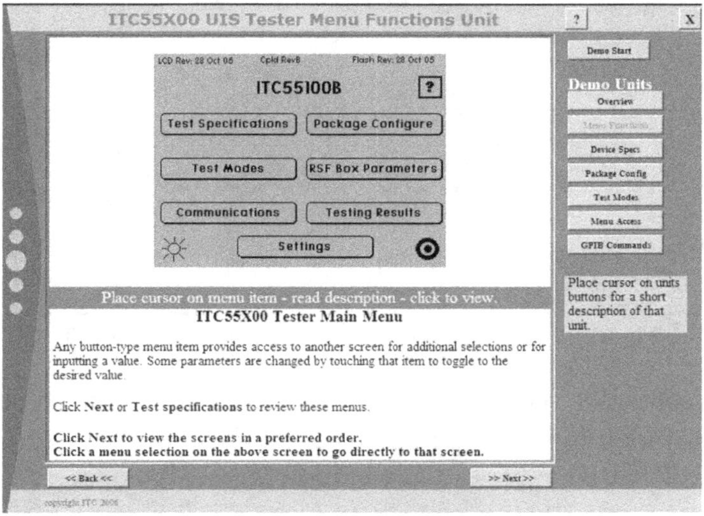

Figure 12–1: Interactive Product Tutorial 1 (800 by 600)

Figure 12-1 shows a simple tutorial that runs on any computer with a browser. It requires the student to click **NEXT** to view the various tutorial modules in sequence. The user can also click on a menu item in the upper screen area and go directly to that function's description.

The student can also click on the separate tutorial modules by clicking the appropriate **Demo Units** button on the right side of the application.

While this tutorial is simple and in its current state does not include audio or video, it does require a student to react.

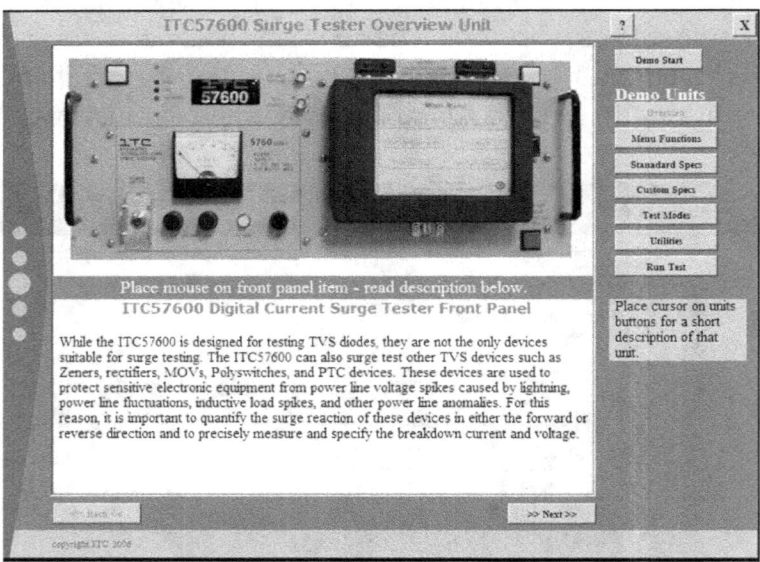

Figure 12–2: Interactive Product Tutorial 2 (800 by 600)

As Figure 12-2 shows, the basic application code can be quickly adapted for tutorials of similar products. The operation is the same as in the first example so a student can quickly understand how to operate and use this tutorial application.

This type tutorial works best for products that use a proprietary user interface that does not run under one of the major operating systems: Windows, Max OS, and perhaps Linux.

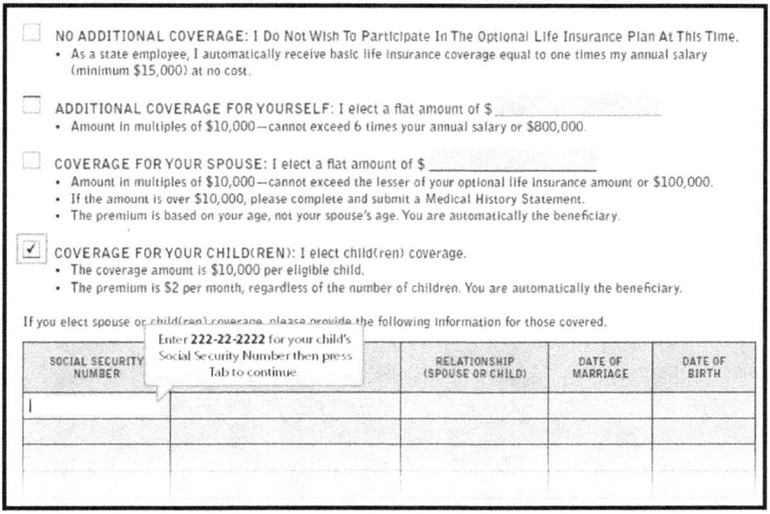

Figure 12–3: Adobe Captivate Tutorial Example

Adobe's Captivate is more that a screen recorder. As the example in Figure 12-3 shows, the student is visually and audibly prompted to click the **Coverage For Children** check box then the **Social Security** number is requested and it must be typed exactly as shown in the prompt box.

You can also use Captivate to show what is to be done by recording the steps, using the screen recorder mode, but the most effective learning method is to make the student perform the actual step before he can proceed to the next step.

This Page Intentionally Blank

Chapter 13

Help Authoring

Help Authoring Overview

HTML Help appears to be the front runner for online documentation at the present time. It is ideal for presenting the user with the most relevant information, regardless of where the user is in the program.

Help Authoring Tools (HATs) deserve their own chapter because these tools are becoming as important as word processors in the technical writing field. Most developers of these tools are trying to provide the user with a "one tool does all".

Once the writer understands this concept, it does make sense. Rather than writing a manual and then copying portions of the manual into the help tool to create appropriate topics, the HAT allows the writer to create the topics and then output those topics in a manual format.

The only problem with this approach is the tendency to write only what is required to satisfy the help calls from an application. This means the writer may not include those chapters or topics that make a manual useful to the user.

This can be remedied by following the guidelines used for creating a manual and by creating those topics to provide the information that would normally be included in the Introduction, Specifications, and Basic Operation chapters of a manual.

Notice that Installation is not included but it could be; however, if installation requires some instructions, installation should be supplied as a separate small manual or brochure. Requiring the help or the product to be operational for reading the installation instructions is counterproductive and could be a major oversight.

Help Tool Features

Most HTML Help authoring tools offer a tri-pane main screen. Some may have four or more panes.

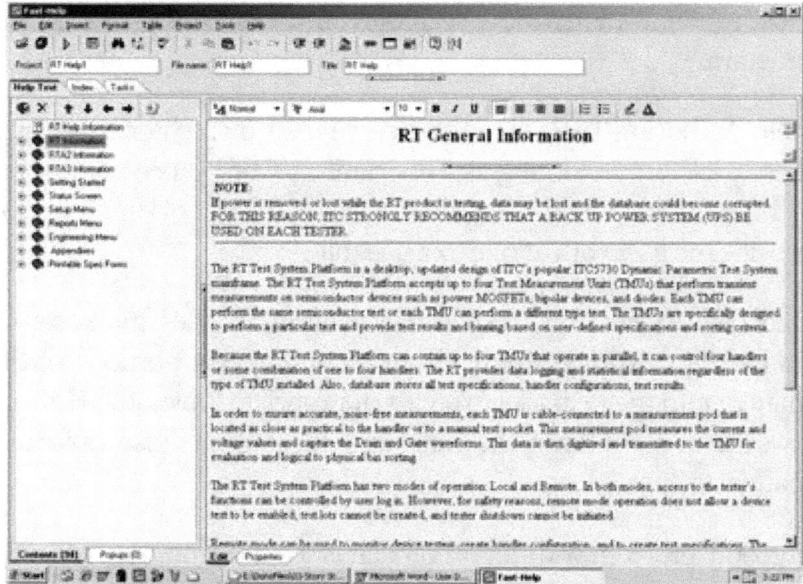

Figure 13–1: Fast-Help HTML Tri-Pane Help Authoring Window

Figure 13-1 is a main screen typical of most HTML Help authoring tools. The small pane at the top is for inputting the header or name of the current topic. The larger pane below the header pane is for inputting that topic's text while the small left-hand pane displays the title of each topic as it is created and assigned.

NOTE: The header title and the title in the Contents pane should be identical. If they are different, the user may be confused when clicking on a title and a topic with a different name appears.

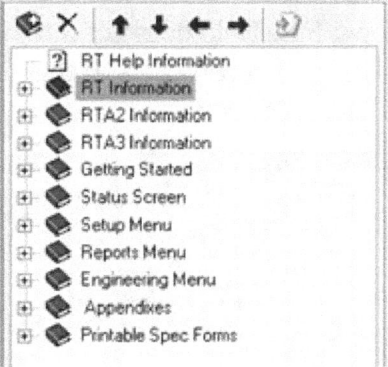

Figure 13–2: Fast-Help Help Contents Pane

Figure 13-2 shows a typical help Contents pane. The topics can be moved up or down and organized into books that are equivalent to the chapters of a manual.

If the intent is to write the help and then output a manual in either a word processor or a PDF format, the organization of the Content pane is very important. Any manual output from the help tool follows the help Contents.

NOTE: Some help authoring tools allow the front and back covers and other manual sections to be included so those divisions are added to a manual's output file but are not part of the compiled help file.

Styles can be created and used by most help authoring tools so that headings and text are styled similar to a word processor. However, HTML Help tools do not allow the complex styles of word processors. This means some things, such as procedure styles, may need modifying for best viewing in a manual.

As Figure 13-3 shows, Fast-Help does not care what type help output is desired until the help is compiled. At that time, the writer can select from seven different options.

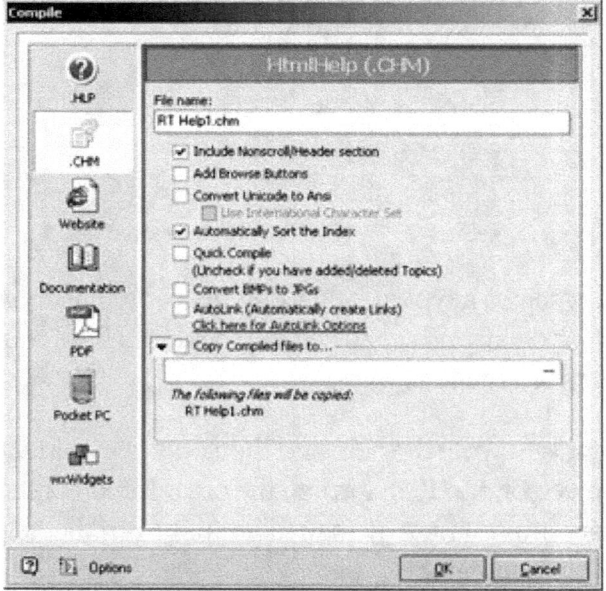

Figure 13–3: Fast-Help Compile Window

Option	Description
.HLP	WinHelp for legacy products (Not supported in MS Vista.)
.CHM	MS HTML Help (Supported in Windows 98 and up.)
Website	Creates a Website help system that can work under Windows and UNIX.
Documentation	Creates a printable document with Table of Contents and Index if specified.
PDF	Creates a PDF file that is readable on most operating systems.
Pocket PC	Creates an .htm file that works on some versions of Pocket PC operating systems. Verify the version based on the help tool.
wxWidgets	A cross-platform framework that supports both HTML Help and its own help format. HTML Help will only work under Windows, but wxWidgets Help will work in Windows, UNIX and on the Apple Mac.

Table 13–1: Fast-Help Compile Options

NOTE: The documentation options apply to both a word processor file and a PDF file output.

Figure 13–4: Fast-Help Document Options

Figure 13-4 shows the options available when Fast-Help is used to compile the help as a document. Most authoring tools offer similar options so the help can be converted to a manual layout. The Table Of Contents (TOC) is automatically created, using the topics as ordered in the help's Contents window.

Each document option displays a screen that either allows the input of text and graphics to create a section of a manual such as Front or End Matter that may not be part of the help system.

NOTE: Some help authoring tools create an eBook executable file that is self-contained. Before buying, decide what features are best for your customers.

When help is compiled as HTML help, Figure 13-5 shows what the Microsoft HTML API displays. This is standard for most Windows operating system and is followed or copied by most non-windows help systems, if they run on a Windows based PC.

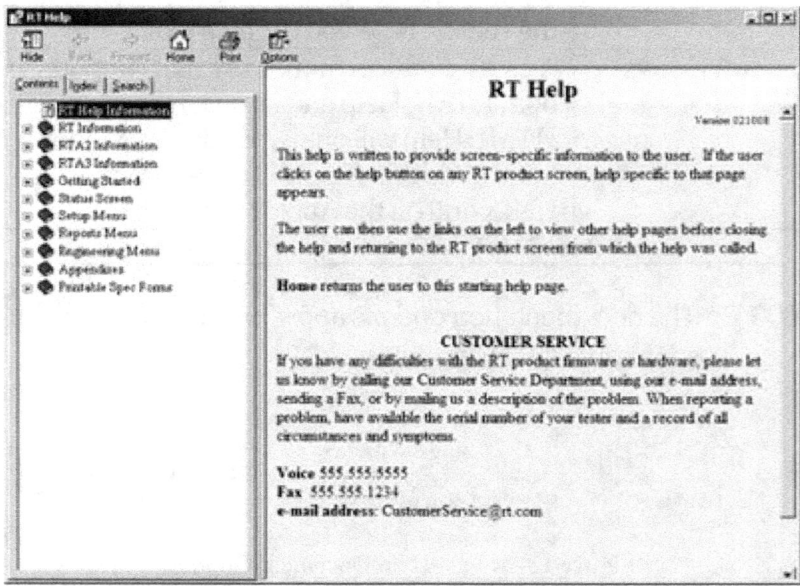

Figure 13–5: Typical HTML Tri-Pane Help Window

NOTE: Use of Fast-Help as examples does not imply an endorsement of the product only a familiarity with it.

Before HTML Help was available, WinHelp was the standard. The best one at that time, in my opinion, was ForeHelp (Figure 13-6), but due to competition it failed the transition from WinHelp to HTML Help. ForeHelp was one of the first to include a topic editor in its program. Some WinHelp authoring tools used Word to create the topic text, which saved those tools development time and money.

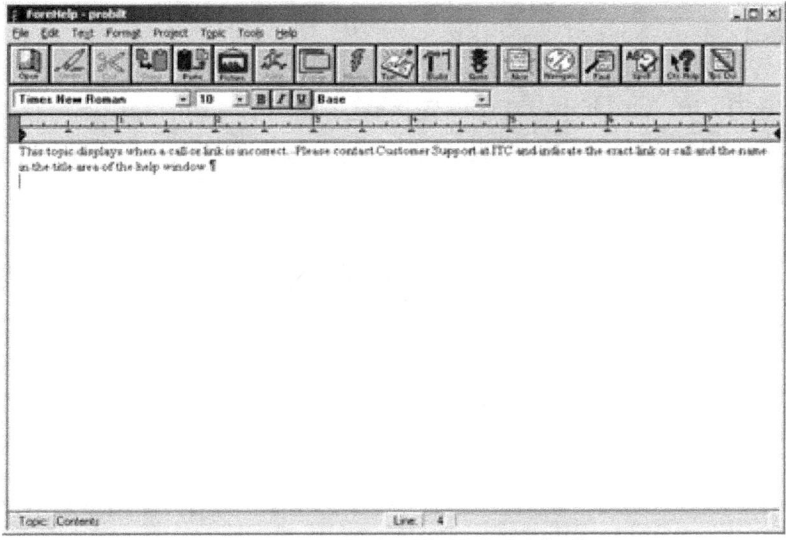

Figure 13–6: ForeHelp WinHelp Main Screen

Notice that ForeHelp was plain looking and things were not as conveniently located as they are on most HTML Help tools, but for its day (pre 2000) it was a great tool.

Most WinHelp tools could import help string files from the programmer, but few HTML tools allow this, so the help writer has to do this manually. For this reason, the programmer must be sure to never change the strings or the help context ID numbers once they are assigned. The more expensive HTML authoring tools may allow the importing of a help string file, but it was common for WinHelp tools.

NOTE: How context strings and IDs are handled may be important when moving from WinHelp to HTML Help.

As with other WinHelp authoring tools, the writer could create special buttons for display in the help window (See Figure 13-7). Some
HTML tools offer this capability but most only allow you to select
from several predefined buttons.

When the WinHelp file is compiled, a window similar to the one
shown in Figure 13-7 is displayed when help is called.

Figure 13–7: Typical WinHelp Topic Window

Help Tool Considerations

There are many HTML Help authoring tools available. Some of the first ones have disappeared or belong to different companies today. So if you are planning to use a help tool for any period of time (3 to 5 years) try to verify the tool's future by reviewing its past.

Before deciding which tool to buy, several things should be considered. This decision should involve programmers, technical writers, training personnel, and management.

> **Author's Aside**
> Most tools allow a time-limited demo version to be downloaded and installed. Unless you have a good registry cleaner, once the tool is installed, the demo cannot be reinstalled after its grace period. This can be a problem if your evaluation period is longer than the demo's grace period or if you want to review a tool after it has been uninstalled.

The following list of things to consider are numbered but are not ordered based on priority or importance.

1. Will topics be created in the tool?
2. Who provides the help strings and IDs?
3. Is help the only desired output?
4. What type help will be created?
5. Will help be one large file on multiple small files?
6. Is cost of the tool a factor?
7. Is tool support important?
8. Will help be translated to other languages?
9. Will video and animation be wanted or needed?
10. Can the tool import or use a CSS file?
11. Do you need to add special help buttons?

These questions are enlarged upon so you can better understand what each question means and what you should look for before committing to a help-authoring tool.

Will topics be created in the tool?
Most HTML Help tools allow the creation and editing of topics in a topic's editing window. Some tools may require the text be created in a word processor and then imported and compiled by the tool.

Who provides the help strings and IDs?
Are the help call strings and context IDs provided to the help writer via a file from the programmer or does the help writer provide that information to the programmer. Can a file be imported to the help tool or must the strings be manually entered?

Is help the only desired output?
Will the tool be used to create a manual or will a manual be written and then help created from the manual? If a manual is needed, will it be outputted as a PDF or as a word processor file? Is an eBook or Pocket PC format output desirable?

What type help will be created?
What type help is required: WinHelp, HTML Help, web help, Net help, etc.? Some tools provide all types of help from one source file while others only provide a couple of output types.

Will help be one large file on multiple small files?
Do your products require one help file for each product or will 4 or more smaller help files be created and combined or merged to appear as one large help file to the user?

Creating a single help file is usually easier but programmers may want separate help files for each major module of an application. This is often the case if a module can be used in several products and the help for that module is slightly modified for each product.

An example might be a User Administration and Log-In module that runs on several products but is coded only once.

Is cost of the tool a factor?
Help authoring tools range in price from about $100 to $1,000 dollars or more. Less expensive tools have fewer functions and output capabilities. They may also not allow for importing text from some word processors or may not export to a file type your word processor can use.

Is tool support important?
Some tools have only e-mail support and some have telephone support but may charge for help after 30 or 90 days. You may also want to consider buying a separate support license if you are not familiar with the tool or with any help tool.

Will help be translated to other languages?
Generally the less expensive tools will not make translation easy. Also, is Unicode supported? This is important for languages such as Chinese or Japanese.

Some tools provide an international version that shows the native language in one window and a second side-by-side window allows the translator to translate that topic. Some tools allow the help to be compiled to each language from the same source file. One source file is much easier to keep up-to-date than multiple source files and to one file makes it easier to keep track of the help changes.

Will video and animation be wanted or needed?
If video and/or animation files might be called from the help, make sure the help tools support these features. Most do but the ease with which they can be created and included varies. Does the tool allow videos to be created or is a separate tool required?

Can the tool import or use a CSS file?

Use of Cascading Style Sheets make help look similar from product to product just as a CSS file controls how a website looks. Some high end authoring tools may allow this but most mid- and low-range tools use styles but do not allow importing of a CSS file.

Do you need to add special help buttons?

Some companies like to add a **Support** button that displays a support or troubleshooting help file. Some may want to add a **Web Help** button to provide additional help. Can the tool do this or does it only allow the addition of pre-selected and pre-named buttons? Can you label the buttons or can you add your own button icons or pictures?

Context Sensitive Help

Help is changing from providing all the help possible to providing only what keeps the user and customer happy. It does appear that most users do not read the help, much like manuals. Users prefer to talk to a person even when the information they seek is readily available as either an online manual or in a help system.

Context Sensitive (CS) help was the norm when WinHelp first arrived on the scene but as time passed, CS help gradually disappeared either because of the additional cost in time and effort or because screen help was considered sufficient.

As shown in Figure 13-8 (next page), when the "**?**" in the upper right-hand corner of the Log In box is clicked and then **User ID** is clicked, context sensitive help for only that field is displayed. This information is also available by clicking the **Help** button but that help describes all items in the box. CS help describes only the clicked item.

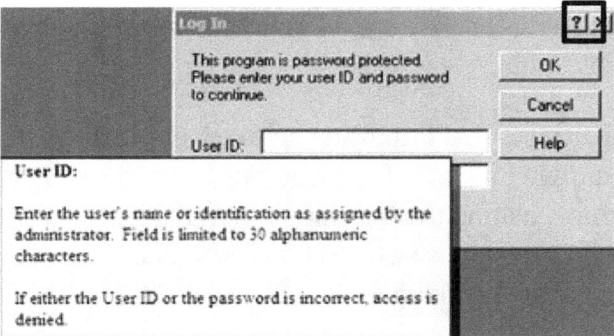

Figure 13–8: Log In Box With CS Help Shown

While CS help is useful for large screens and dialog boxes that have 8 or more items, it is not that useful for small dialog boxes such as the Log In box shown in Figure 13-8. Therefore, CS help is being phased out in favor of screen or dialog box help. Truth be told, CS help is a lot more work, not only for the help writer but also for the application programmer.

Starting A Help Project

Some of you may be wondering why the fuss about how help strings are created or used. Over the years, I have found it was better and easier for the programmer to provide the help call strings to the help writer rather than the other way around.

NOTE: Some help tools say they can scan a program and automatically create the help strings. I have not been tempted to try this method, but it may be worth looking into.

If the application is small and has only 100 help strings, manually typing those strings into a help tool is not a major chore. However, if the application has several hundred to several thousand strings, it becomes a major effort both in time and cost, especially if the programmer renumbers the help strings when making major updates to the application.

Also, I have found that importing help strings into the help tool and creating a topic for each string that has the name of that string is quicker than creating those strings when that help topic is written. When writing help, it is easy to go to a screen or dialog box in that application and click its help button. When the button is clicked, and assuming the help and the application are working correctly, the topic string associated with that help button is displayed.

I can then find that string in the help tool, write the help for that screen or box, rename the topic so it makes sense to a user and that topic is done. While this may not be the most efficient method, it does work well, especially for ensuring that the correct help is displayed for each help call from the application.

To ensure the correct help context number is called (programmers can occasionally make a mistake), I recommend Microsoft's Help Workshop for monitoring each help call from the application. Help Workshop displays the string and context ID number actually called, verifying that the topic help is correct for that help button. This check can be done if the programmer has labeled the strings with the screen or dialog box name that calls the help topic.

NOTE: Help Workshop is a free tool from Microsoft.

Sometimes the help context ID number may change because the programmer has to make major changes or a mistake was made. This is where the ability to import an HM or a similar file supplied by the programmer comes in handy so the topics can be renumbered. While it is true, the help writer could supply a file to the programmer for importing, it appears to work better the other way around.

This renumbering of context IDs is not as simple as importing a new string file, even in the old WinHelp tools. I found a work-around that does the trick but some HTML Help tools do not allow a file to be imported, so back to square one.

Call strings have some naming conventions that must be observed. Check with the programmer for these conventions. Usually HID strings are for menu and icon help; HIDD strings are for help buttons; HIDC strings are for context sensitive help calls; HIDP strings are usually for interface calls and are not always used but may be included by the programmer.

Help Call String	Context ID #
HID_HT_NOWHERE-----------------------	0x40000
HID_HT_CAPTION-----------------------	0x40002
HID_HT_SIZE--------------------------	0x40004
HID_HT_HSCROLL-----------------------	0x40006
HID_HT_VSCROLL-----------------------	0x40007
HID_HT_MINBUTTON---------------------	0x40008

Figure 13–9: Typical Help Strings From A Programmer

Help Workshop

Help Workshop is a free help tool from Microsoft and usually comes with most help tools. It is a handy tool for tracking help calls from an application to its help file.

This tool, while primitive in nature, can compile and create a help file using files (.rtf) exported from a word processor. It is not that intuitive or easy but not bad for a free tool. Unless this is the only tool you can afford, I suggest you buy one that is easier to use and has more features.

Figure 13–10: Help Workshop's Main Screen

When Help Workshop is activated, the above screen is displayed. The **Help** menu offers some useful information on using the tool. There is a training card help feature that calls a small WinHelp screen that steps the user through the creation of a project.

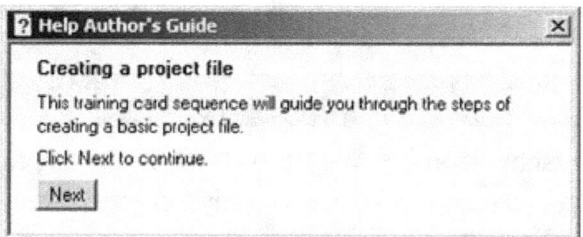

Figure 13–11: Help Workshop's Training Card Help

When Training Card is clicked on the **Help** menu, the above help is displayed. What makes this neat is that when the instructions are accomplished in the screens that follow after **Next** is clicked, text changes to provide the next step in the sequence.

However, the only thing I use Help Workshop for is to display the help calls made from an application to that application's help file.

To display these help calls, click **View** and then click **WinHelp Messages** to display the Messages screen.

Figure 13–12: Help Workshop's View Menu

The Message screen shown in Figure 13-13 displays the help calls made by an application to its help file when a Help button is clicked in that program.

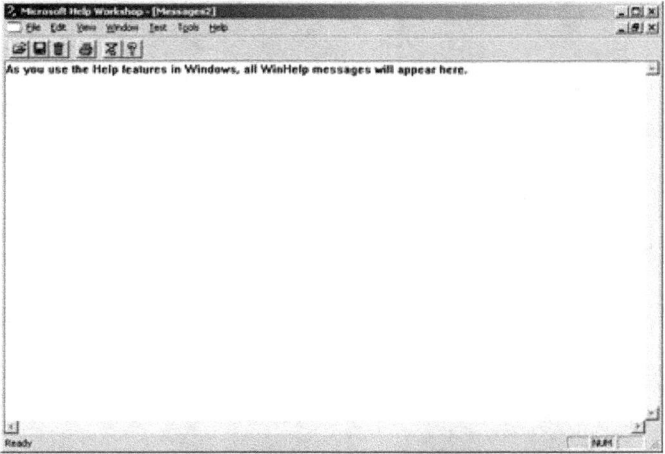

Figure 13–13: Help Workshop's Message Screen

NOTE: There is also an HTML Help version and it works much the same way as the WinHelp version.

The HTML version of Help Workshop does not have the Training Cards help, which indicates that WinHelp still can do some things that HTML either cannot do or can only do with special scripting.

Help Examples

The following are examples of some help available to a user, depending on the product. There is no consistent help standard but appears to depend either on a company's policy or the help writer's creativity. Larger companies can afford to develop their own help systems that show their products in the best light.

If a company can afford to develop its own help system, that appears to be the best long-term solution. Help authoring tools do not appear to have a long lifespan even when a large company owns the tool unless that company uses that tool for its products' help.

Figure 13–14: Microsoft FrontPage 2002 Help

Microsoft FrontPage 2002 (Figure 13-14) help appears as a standard Dual-Pane help system except for the **Answer Wizard** tab. I am not sure if this tab can be added, using a third party tool, but I believe Microsoft provides a way to do this, or at least did at one time.

Almost all applications have help included that is accessed by clicking the **Help** menu on the top-level menu. The HTML Dual-Pane help window is a standard for Microsoft windows, but it appears Microsoft uses a proprietary help authoring system for some of their high-end programming products.

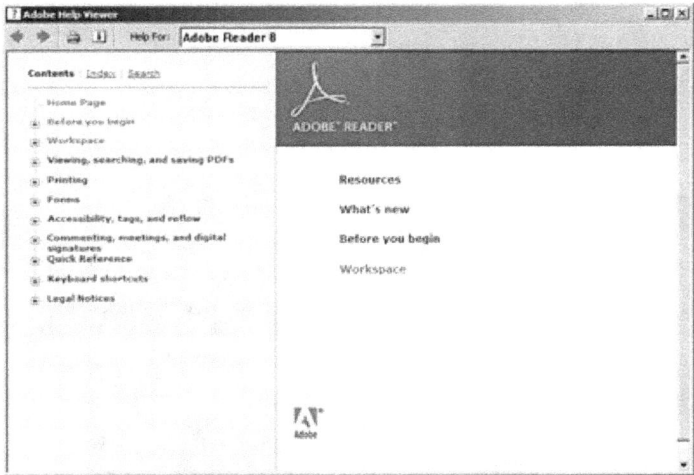

Figure 13–15: Adobe Reader Help PDF Viewer

Figure 13-15 is a special PDF viewer for Adobe's Readers. Adobe favors PDF files to present help for its Acrobat products. They also use browser-based HTML plus online web help. However, webhelp only works if the user has access to the Internet. This approach is not good for test equipment and other production type products because companies try to limit Internet access to reduce employee temptations.

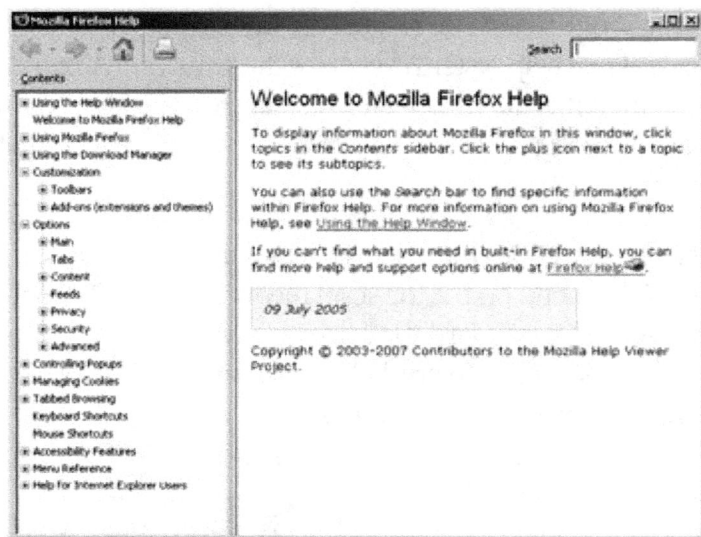

Figure 13–16: Mozilla Firefox Browser Based Help Screen

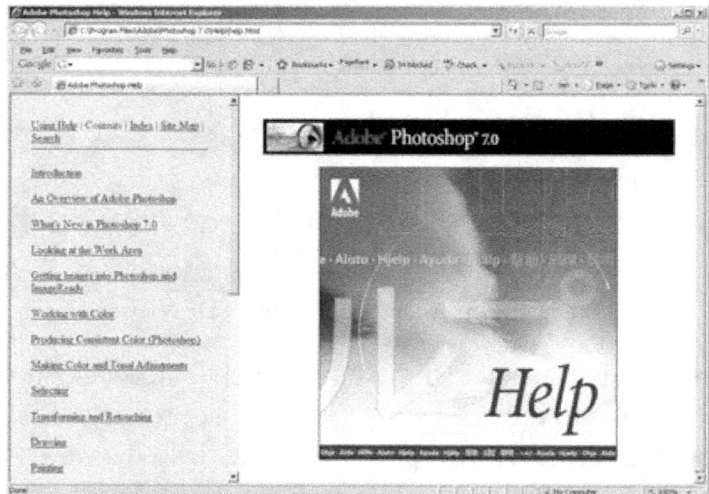

Figure 13–17: Adobe Photoshop Browser Based Help Screen

The help concept in Figure 13-18 was developed for use on a tester with a touch screen. Buttons were initially used because I assumed they were easier to finger-touch than normal HTML links. Turns out that a pointer was used and buttons were not needed.

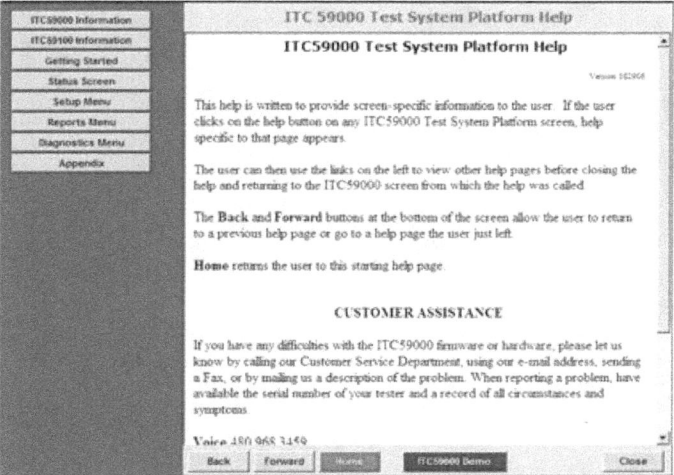

Figure 13–18: Touch Screen HTML Help Concept

This help concept uses iframes and CSS positioning to place the iframes in the desired location. I did not use <DIV>s because of the need to call content from one frame to be placed into another iframe. Actually, I replace the contents of both iframes when any button is touched or clicked in the navigation frame.

This Page Intentionally Blank

Chapter 14

Editing The Words

Editing Overview

Before giving the manual to someone else to proofread, read it several times. Some writers are not suited for correcting their own work; their personalities are such that once the words are written, they assume the job is done. If you are someone who hates proofing, find someone who enjoys it. Thankfully, there are such people.

Proofing and editing require concentration and they can be exhausting. Editing is the process of making sure that what was written is what was meant and that the words are suited for the intended audience. Editing is determining not just if the **correct** word was used but if the **best** word was used.

Editing is also determining if the words make sense in the order they appear in the document. Editing is not only verifying that all words are spelled correctly, but also that they are used correctly.

Editing Tips

Here are some tips that should help when editing. Remember editing is not only done once the manual is finished but should also be an on-going process. Minor editing occurs during writing, major editing only after you have finished the first or second draft.

Here are some things I have learned the hard way over the years and few I have recently learned.

1. Keep spell checker on while writing.
2. Let the writing cool before editing.
3. Reread quickly for sense of flow.
4. Change font size and page view.
5. Use the grammar checker.
6. Print document to paper.
7. Read text out loud.
8. Read last sentence first.
9. Create TOC and check its order.
10. Give to someone else to edit.

Keep spell checker on while writing
Some writers prefer to turn off the spell checker until it is time to proof or edit. If you are a superb speller and an accurate typist, this works well; otherwise, correcting misspelled words as they are typed saves time and improves your spelling.

Read quickly for sense of flow
Read the manual from beginning to end or one chapter at a time to get a feel for its flow. This quick read could uncover areas that are not well written or things that are not properly described. At this point mistakes are secondary to the overall sense of the manual.

Let writing cool before proofing and editing

When proofing the manual, do not start until the writing has cooled. It is easy to overlook errors because you know how you write. After you have written for a few years, you develop a certain rhythm and style that becomes subconscious. This means you tend to see what was intended instead of what was actually written.

Change font size and page view

Another aid is to change the look and feel of the manual. Changing the font size or changing the document's view from 100% to 150% changes the look and the feel. Switching to a layout or normal view instead of the page view makes the manual appear different and helps you to see errors better.

Use the grammar checker

Turn on the grammar checker. Word includes a grammar checker and while it may not be the best, it helps focus your attention on one sentence at a time. Grammar checkers can be more of a nuisance than a help, but using one allows you to see what the checker objects to, usually in a separate window. This one-sentence view forces you to concentrate on that sentence. **Don't assume the grammar checker is always correct!**

Print document

Now that you have made all the corrections you can find, print the complete manual or one chapter at a time. Over the years, I have found detecting errors is often easier when the manual is read from paper rather than from a monitor.

Read text out loud

Find a quiet place and read the manual out loud. Notice when you pause. Is there a comma or some other punctuation there? If not, should there be? Most writing is improved by punctuating it as if you were talking to someone. Also, reading out loud helps to spot words and phrases that slow or cause the reader to misunderstand what is written. If you have to stop to understand what was meant, that area should be rewritten.

Read last sentence first

Some editors suggest reading the manual in the reverse order. That is, read the last sentence or paragraph first then the next to the last and so on until the first sentence or paragraph is read. This technique forces you to concentrate on the actual words instead of the intent. The only difficulty with this method is sometimes one sentence or paragraph depends on a previous sentence or paragraph for its meaning, but the sentence should still make sense.

Create TOC and check its order

As a last check, create or update the Table Of Contents and review its order. Often this shows if things are described and discussed in the proper order. Also, at this stage, you can see if the paragraphs follow the intended outline and if not, why not. Check that you have Headings in the correct order and that you do not have a Heading 3 without first having a Heading 2.

Give to someone else to edit

Before turning in the manual, give it to someone that has good spelling and grammar abilities. This person does not need to fully understand the subject, but he may spot things the writer missed. Have this reader circle anything that bothers him. The writer can review and decide if something needs to be rewritten or if a different word should be used, depending on the intended audience.

Evaluating Documents

Once you have created your user document, whether it is a manual, a video, a tutorial, or a marketing brochure, you should attempt to evaluate how effective that document is.

As a rule, the person that developed or designed the document should not be the sole judge of how well that document works nor should his immediate superior. It is best to use a person that is interested only in the results of that document and who did not have any real part in its development.

Of course, the best judge of any document is the intended audience. After reading a manual, watching a tutorial, or hearing a procedure described, can that user operate the product or perform that procedure. If not, either the document is inadequate or the user has no incentive to learn. Determining which is the problem, requires an impartial judge.

While quiz testing is popular and often a good way to discover how well a user absorbed a document, some people freeze when presented with a test. Also, some people can memorize the answers but not understand them well enough to use that learning in real-world situations. The best indication of learning is the application of the learned information. This can be done on the actual product or on a simulation of that product. A simulator is safer—harder to damage.

When a course or tutorial is taught, the instructor should not determine the effectiveness of that course. The instructor has a vested interest in showing that the course works, especially if the instructor also had a major part in creating that course.

This Page Intentionally Blank

Appendix A

Frequently Misused Words

Here is a list of words that seem to cause writers the most trouble. Some are often misspelled and some are incorrectly used because the words are similar in spelling but different in meaning.

Words	Best Use or Meaning
accept	means to receive
except	means to leave out
adapter	a person who adapts
adaptor	device, thing, connector
advice	noun
advise	verb
aid, aids	means to help someone (He aided her.)
aide, aides	refers to the person helping (He was her aide.)
affect	influence/change - usually a verb except in Physiology
effect	N = result; V= accomplish
alright	use all right (better = agreed or OK)
already	already left
all right	all (set, packed) ready to go
among	more than two involved
between	more than two but thought of individually
amoral	questions of morality
immoral	evil or bad
among	more than two involved
between	more than two but thought of individually

Words	Best Use or Meaning
amount	cannot count
number	can count
anybody	both are one word means any person
anyone	means any single (one) person
axel	ice skating jump
axle	rod between wheels
bases	four bases in baseball
basis	claim has no basis
bait	for catching fish
bate	bated breath
born	babies are born
borne	loads are borne
but	but why?
butt	a goat butted me.
bus	a bus stop
buss	a kiss
can	implies ability
may	implies permission
canvas	a fabric
canvass	to solicit
capital	upper-case letter
capitol	a building
chafe	make sore by rubbing
chaff	to tease
chord	musical notes
cord	a rope
course	of course, golf courses
coarse	not fine - a texture
compare to	liken similar items
compare with	to illustrate similarities or differences
complete	an absolute – not more complete
	most complete and less complete do work
country	geographical
nation	political

Words	Best Use or Meaning
criteria	plural
criterion	singular
data	as facts = plural; as information = singular
device	noun
devise	verb
different than	use when clause follows
different from	use to compare two things
discreet	good judgment
discrete	separate parts
Earth	the planet
earth	Dirt or soil
farther	indicates distance
further	indicates time , quantity, additionally (furthermore)
finalize	avoid using - reword
firstly	use first also second and third no "ly"
flout	mock, scoff, distain, disregard
flaunt	show off
forgo	do without
forego	to precede
forth	to go out from
fourth	a number
good	good dog (adjective)
well	dog is well (adverb)
hertz	frequency
hurts	pain
imply	suggest
infer	deduce
insure	insure your car
ensure	to make sure
assure	to give confidence
intend	signify or mean (verb)
intent	aim or purpose (adj.)
importantly	rephrase - avoid use
irregardless	use regardless

Words	Best Use or Meaning
it's	it is
its	its mine - possessive
joule	unit of energy
jewel	a gem stone
less	quantity - count unknown
fewer	number - can be counted
like	preposition, used to compare things
as if	followed by a verb - conjunction, use before a clause
lightning	a bolt from the sky
lightening	to make less dark
load	a weight
lode	a vein of mineral
loath	adjective = reluctant
loathe	verb = to despise
loose	not solid
lose	not found
metal	a material = gold, copper
mettle	courage or spirit
militate	operate against
mitigate	to assuage or soften
pail	a small bucker
pale	lacking color
parlay	to gain
parley	to discuss
pedal	pedal on a bike
peddle	to sell goods
perform	accomplish
preform	shape beforehand
prescribe	rule or guide
proscribe	prohibit
principal	chief person or thing
principle	doctrine, law
prophecy	noun
prophesy	verb

Words	Best Use or Meaning
recent	new
resent	dislike
shall	for first person, will for second and third person
should	ought to do
will	desire or wish or probability
unique	do not use with more - unique is absolute
utilize	prefer use
vice	a bad habit
vise	a tool
want	desire
wont	habit or custom
which	use with unneeded clauses (non-restrictive)
that	use with needed clauses (restrictive)

This Page Intentionally Blank

Appendix B

Transitional Words & Phrases

Transitions between ideas can be confusing. Here is a list of commonly use transitions and how they should be use based on the clauses they separate.

Transitions		Best Use
and additionally also	indeed in fact	joins ideas of equal weight
for instance for example	for one thing	illustrate or expand a point or idea
similarly	likewise	comparisons of ideas
therefore consequently	so then	supporting clause backs up main clause
finally all in all	on the whole in short	summarizes main clause
of course no doubt certainly	to be sure granted that	concedes main clause point or diverges from main clause point or idea
but yet	however	reverses main clause idea and returns to original point
still nevertheless		returns to main clause thought or point after conceding a point
although	though	adds a concession to a point
because for	since	adds a reason to a point or idea Use since only to indicate time

Transitions		Best Use
if unless provided that in case when		use to qualify or restrict a point or idea
as if as though		makes point or idea conditional or tentative
when while soon after before until		use to temporary relate points and ideas.

Appendix C

Common Abbreviations

Technical writing requires the use of abbreviations more so than most other types of writing. When an abbreviation is used, spell out the full term the first time it is used unless it is so common as to not require an explanation for the intended audience.

Abbr.	Term	Abbr.	Term
A	ambient	A	ampere
A/D	analog to digital	ac	alternating current
Ah	ampere-hour	B	bel
b	bit	b/s	bit per second
B	byte	Btu	British thermal unit
C	coulomb	CPU	Central Processing Unit
D/A	digital to analog	dB	decibel
dBm	decibel in milliwatt	dc	direct current
F	Farad	G	ground
GHz	gigahertz	gnd	ground
H	Henry	Hz	hertz
IPS	instructions per second	ips	inches per second
K	Kelvin or kilo (1,024)	k	kilo (1,000)
KB	kilobyte	Kb	kilobit
kV	kilovolt	kW	kilowatt
LSB	least significant bit	M	mega (million) or 1,048,576

Abbr.	Term	Abbr.	Term
m	meter or milli (thousandth)	mA	milliamp
MB	megabyte	Mb	megabit
mF	millifarad	mH	millihenry
MHz	megahertz	mm	millimeters
MBS	most significant bit	mV	millivolt
mW	milliwatt	MW	megawatt
MΩ	megaohms	mΩ	milliohms
nA	nanoampere	nF	nanofarad
ns	nanosecond	nΩ	nanoohms
nW	nanowatt	p	pico (trillionth)
p-p	peak to peak	pA	picoampere
pF	picofarad	ps	picosecond
pΩ	picoohms	psi	pounds per square inch
pW	picowatt	rms	root mean square
s	second	sq	square
UHF	ultra high frequency	V	volt
Vac	volts alternating current	Vdc	volts direct current
VHF	very high frequency	W	watt

Symbol Table

Symbols			
Ω	ohms	°C	degree Celsius
°F	degree Fahrenheit	°K	degree Kelvin
μF	microfarad	μs	microsecond
μΩ	microohm	>	greater than
<	less than	x or *	multiply by
÷ or /	divided by	|	separator Y|N

Appendix D

Reference Materials

Writer's References

A technical writer, no matter how good, may need to look up some "best usage" rules. It is a good idea to have the following books or similar ones close at hand.

Company Style Manual (if available - if not start one)

The Chicago Manual Of Style

Webster's New Collegiate Dictionary

Microsoft The Manual Of Style (Electronic/software terms)

Strunk &White The Elements of Style

Webster's Standard American Style Manual (Out of print)

Some Links

http://www.writerswrite.com/technical/

http://www.klariti.com/technical-writing/index.shtml

http://www.ent.ohiou.edu/~valy/techwrite.html

http://www.docsymmetry.com/

http://www.coping.org/write/improvtech.htm

http://lorien.ncl.ac.uk/ming/Dept/Tips/writing/writeindex.htm

http://www.christianwritersinfo.net/tips3.htm

http://www.fast-help.com/

http://www.ebookstarter.com/

http://www.ebookresourcecenter.com:80/

http://madcapsoftware.com/products/flare/

http://www.innovasys.com/products/hs3/overview.aspx

http://en.wikipedia.org/wiki/Help_authoring_tool

http://www.adobe.com/products/robohelp/

http://www.drexplain.com/

Appendix E

Manual Cover Examples

Manual Covers

Here are some examples of manual covers. The idea of a cover is to identify the product without confusing the user.

Notice the product's name is in small letter.
Must not be important?

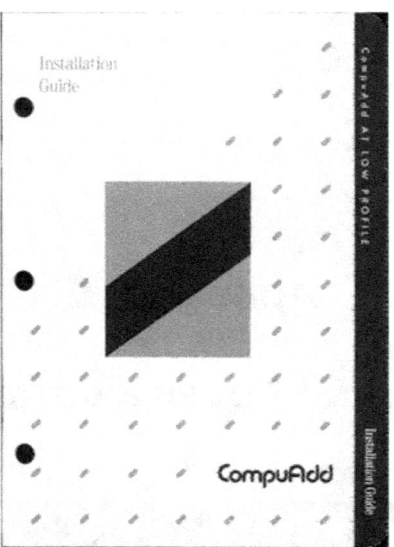

This is an Installation Guide but for what product?

Well identified guide and product.